AF345657

Bilayer's Potential: Electrons Unlock Properties

Ravi

Copyright © 2024 by Ravi

All rights reserved. No part of this book may be reproduced in any man-ner whatsoever without written permission except in the case of brief quotations embodied in critical articles and reviews.
First Printing, 2024

Table of Contents

Chapter 1. Introduction

1.1 Introduction

Throughout recent years, researchers have been showing significant interests in bilayer graphene (BLG) owing to its electrical field-dependent tunable band-gap properties [1-5], which point towards the potential for enormous electronic applications. Furthermore, BLG possesses many extraordinary properties like single-layer graphene (SLG), including high electrical conductivity [6], high thermal conductivity of about 2800 $Wm^{-1}K^{-1}$ at room temperature [7-8], high transparency with white light transmittance [9] and high mechanical flexibility and stiffness [10-11]. Therefore, BLG has become promising for applications in high-frequency transistors [12], flexible electronics [13], photonic devices [14-15], thermoelectric devices [16] and also in batteries for energy storage [17-18]. However, it is crucial to relate electronic and lattice interactions, i.e. electron-phonon coupling which can aid to realize material properties, for instance- formation of polarons, superconductivity, the Peierls transition and Kohn anomaly [19]. The study on vibrational properties of the graphene system has become influential during the last few years [20-26] since electron-phonon interactions, thermal conductivity, optoelectronic properties are notably dependent on the phonons in graphene flakes. Moreover, by probing the phonon properties of the graphene system, it is possible to get a detailed overview of interlayer interactions, stacking order, number of layers and the presence of any defects [27].

Recent breakthroughs in Chemical Vapor Deposition (CVD) technique has opened up an era of exploiting ^{12}C and ^{13}C atoms in either layer of bilayer graphene flakes. BLG with hybrid layers of ^{12}C and ^{13}C atoms can show diverse phonon properties while comprising the same electronic properties as in pristine layers of ^{12}C atoms [28]. In addition, during the fabrication process of BLG, it is very frequent to find vacancies [31]. In the vacancy disordered graphene system, it is possible to observe stable interlayer bonding owing to its neighboring layers' vacancies and lower coordinated atomic sites [32]. Moreover, the presence of a small number of vacancies in the graphene system has the ability to change the low-temperature heat conduction properties. While there is no significant impact on electronic properties due to isotopes, vacancies have notable effects on electronic properties in graphene system [33]. Thus, electron-phonon coupling in the graphene system can be altered significantly in the presence of vacancies.

The presence of defects in the sp^2 bonded C-C structure in BLG can lead to strong changes in phonon properties. When defects are present, translation symmetry of the graphene system can be disrupted which in turn leads to the disintegration of momentum conservation through stimulating interior Brillouin zone K-point phonons. Consequently, Raman active D peak can be triggered in disordered samples [34-36]. Phonon modes can be distributed randomly in any disordered system instead of known frequency regime of pristine graphene [37-38], which can be termed as localized phonon modes alike Anderson's localization of electrons. Mean free path of localized phonons become finite and proportional to the square of the localization length. Since, phonons can be scattered at the presence of vacancies or ^{13}C isotopes in BLG, their lifetimes and mean free paths get confined into finite values. Therefore, it is crucial to study the combined effect of isotopes and vacancies in BLG.

Moreover, during the transfer of CVD graphene to any substrate, it is not possible to avoid wrinkles which might change the phonon properties of graphene [39]. Different types of substrate can induce different amount and type of strain in the system. And strain can induce shifts in the vibrational frequencies of the Raman active bands in graphene. Phonon softening can take place at the presence of tensile strain, while compression might induce phonon hardening [39].

Since low-frequency regime acoustic mode phonons have higher group velocities, they play a significant role in heat conduction in graphene system. Besides, shorter wavelength regime optical mode phonons are greatly affected by the isotopes and vacancies and become immobile. Thus, thermal conductivity is greatly affected by the high-frequency region localized optical mode phonons at the presence of isotopes and vacancies. This might occur due to the scattered phonons originated from mass mismatching in isotope disordered system. Furthermore, by taking consideration of the localization length, former study [37] showed that isotope disordered system exhibits mainly three transport regimes (localized, diffusive and ballistic) owing to the wave interference formed by isotopes. However, Savic et. al. [37] pointed out that it is difficult to observe the localized modes as thermal conductivity is mostly supported by diffusive and ballistic phonons. Since high-energy optical mode phonons exhibit localization effects, appropriate modeling is needed to probe the localized phonon modes in BLG.

1.2 Types of Defects in Bilayer Graphene

As discussed earlier, during the mass fabrication of graphene, defects are expected in the sample. Defects are responsible for breaking the symmetry of the infinite honeycomb lattice of BLG. Different types of defects are possible, for instance, isotopes, vacancies, edges, implanted atoms, doping and so on. Carbon hybridization can also take place in BLG; such as from sp^2 to sp^3 hybridization.

1.2.1 Vacancy defects

Vacancies are very common in BLG. This type of defect greatly modifies the electronic properties of graphene. By studying the electronic inverse participation ratio Pereira et. al. suggested that localized states are induced by vacancies with responsible resonant peaks at Dirac point [33]. Those states are responsible for the localization around the Dirac point. During any substitutional doping, there might be correlative effects between vacancies and dopant atoms [40-41]. By employing Density Functional Theory (DFT), Hou et. al. studied the total energy of graphene at the presence of N dopants [42]. They found that N dopant atoms and vacancy sites attract each other, which in turn suggests that N dopant can lead to more vacant sites or vice versa. This interaction between N dopants and vacancies can alter the role of N atoms for free carrier production in the bulk π bands.

1.2.2 Isotope defects

In any natural carbon-based samples, 98.9% are ^{12}C atoms while only 1.1% are ^{13}C atoms. At the presence of isotope disorder in graphene system, properties like thermal conductivity [43], electron-phonon coupling [21] can be modified owing to the change in average atomic mass. Phonon frequencies are usually supported by the mass difference of isotopes in any isotope disordered system [44]. Thus, it is possible to observe the effects of isotopic mass variation by spectroscopic tools like Raman spectroscopy, electron energy loss spectroscopy (EELS), infrared (IR) spectroscopy and so on. If we take the intrinsic spectral linewidth of Raman spectroscopy into consideration, it can be observed that isotopes play an important role in intrinsic scattering mechanisms. By inducing ^{13}C isotopes in any graphene-based materials, it is possible to analyze the phonon properties [45] as well as the growth mechanism of that sample [46]. Since isotope enrichment doesn't alter the electronic properties of graphene-like materials, the difference between electron-phonon interactions and electron-electron interactions can be sorted out very easily [45]. Moreover, when ^{13}C atoms are induced in any ^{12}C based gas source, it becomes easier to study the crystal growth of carbon nanotubes [47] or graphene [46] from that gas source.

1.2.3 Edge defects

In honeycomb lattice of graphene-based materials, two types of edges can be observed- armchair and zigzag edges as shown in Fig. 1.1 for SLG.

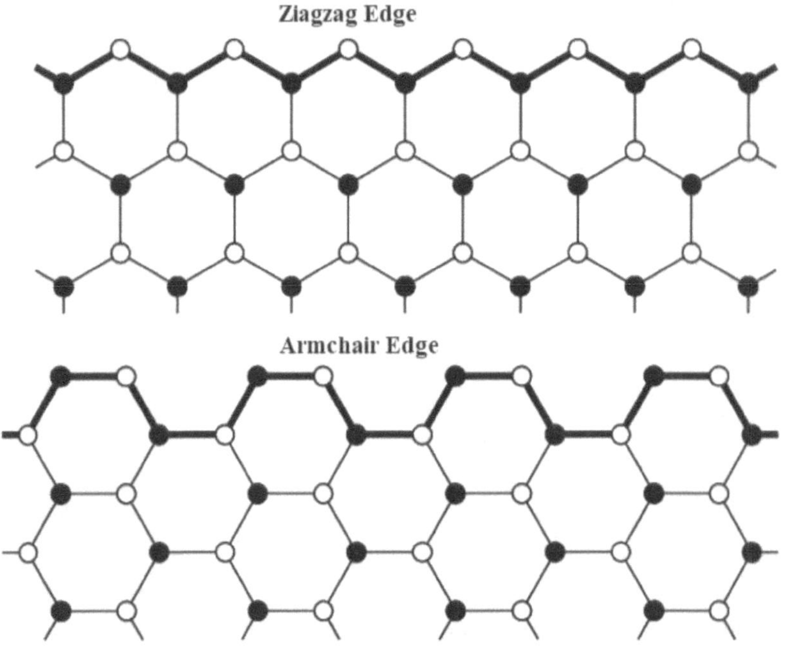

Fig. 1.1. Different type of edges of SLG.

The edges of graphene have a greater impact on the phonon and electron transport properties, especially in nanoribbon structured graphene samples. Edges play a vital role in the low energy spectrum of π electrons [48]. Furthermore, graphene nanoribbons can show a bandgap for edge prompted chirality [49-50] as well as certain superconductive [51] and optical properties [52]. In addition, the magnetic properties of nanographene samples are largely dependent on the edges owing to their notable contribution to the Fermi level density of states [53]. Thus, for graphene nanoribbon, it is a crucial need to investigate the effect of edges on the phonon properties.

1.3 Motivation

Owing to its fascinating thermal and electronic properties, two-dimensional structured graphene has enthralled many researchers for the last few years. To be more specific, bilayer graphene exhibits high electron mobility and it has the electrical field-dependent bandgap tuning feature to be used as a switch-off electrical conductor. Regardless of those properties, it is impractical to ignore the fact that achieving a pristine graphene sheet is very challenging. Structural defects are very usual while producing graphene sheets. Yet it is also feasible to acquire some novel properties by exploiting defects.

In recent years, many computational models have been employed to quantify the phonon properties of BLG including atomistic finite element (atomistic FE) model [22], Born-von-Karman (BvK) model [23], extended tight-binding model [24] and density functional theory [25], yet majority of them focused only on pristine BLG. Furthermore, some experimental works have been done on the phonon properties of BLG considering isotope effects [21, 45]. The effect of combined isotopes and vacancies on the phonon properties of hexagonal BN has been pointed out in previous work [54]. However, as far as my knowledge goes, there has not been any work by acknowledging the combined effect of isotopes and vacancies in BLG. The regular topological symmetry is disrupted by the presence of any defects and it leads to complex modeling of the lattice system. Thus, conventional dynamic matrix-based techniques become inappropriate since they need an enormous amount of time and memory space for computation. On the other hand, a model should be sufficiently large to extract the phonon modes of defect induced system. Therefore, a new model is imperative to realize the vibrational properties of isotope and vacancy induced bilayer graphene.

1.4 Objectives

The aim of this work is to study the effect of different types of disorders in the BLG sample. Forced vibrational method [55] has been employed to estimate the phonon properties of BLG and its nanostructured ribbon. This method can be helpful to calculate the phonon modes of a large disordered system by a reduced amount of time. Moreover, the phonon density of states (PDOS) of any system can be calculated accurately at a very low-frequency regime. In addition, PDOS can be calculated at any certain resonant frequency.

At the presence of any disorder in any BLG sample, phonon localization can take place which hinders the phonon transport properties. Thus, it gives rise to reduced phonon lifetime. In any disordered sample, the mean free path of phonons becomes definable which is proportional to the square of localization length. In spite of having higher thermal conductivity, defects in any BLG sample mitigate the phonon transport owing to the deformation of the pristine crystal structure. Since localization hinders the mobility of phonons, they can no longer act as heat carriers. Therefore, it is of crucial importance to study the effect of phonon localization in BLG. In this work, especially the following objectives have been accomplished-

1) To develop a sufficiently large and definitive model by employing the FV method to study the phonon properties of BLG.
2) To estimate the combined isotope and vacancy effects on the phonon modes of BLG for the first time.
3) To analyze the vacancy and edge effects on the vibrational properties of BLG nanoribbon.
4) To compute the thermal conductivity of different BLG nanoribbon system at the presence of disorders.

1.5 Contribution and Achievements

This book contributes to the study of phonon properties and thermal conductivities of different BLG samples.

The contribution of the author includes-

1) For the first time, a model of large scale BLG has been obtained to compute the PDOS at the presence of different types of defects.
2) FV method has been employed to simulate the localization length and localized phonon modes of BLG. This contribution will be helpful to analyze the heat conduction properties of BLG based devices.
3) Effect of edges, as well as vacancies on the thermal conductivities of armchair edge bilayer graphene nanoribbon, have been studied thoroughly.

4) Raman active G peak has been estimated for different isotope enriched BLG samples which are very expensive to obtain experimentally.

The results demonstrated in this book were achieved by the author during the period of study at the University of Ottawa to fulfill the requirement of pursuing an M.A.Sc. The work was accomplished under the supervision of Professor Jeongwon Park at Park's Lab of the University of Ottawa from Fall 2018 to Winter 2020. The following publications related to this work have been submitted during this period.

1) **Khalid N. Anindya**, Md. Sherajul Islam, Jeongwon Park, Ashraful G. Bhuiyan, Akihiro Hashimoto. " Interlayer isotope and vacancy effects on the phonon modes in AB stacked bilayer graphene." *Carbon*, 2020 (Submitted, Under review).
2) **Khalid N. Anindya**, Md. Sherajul Islam, Jeongwon Park, Ashraful G. Bhuiyan, Akihiro Hashimoto. "Impact of vacancy defects on the phonon modes in AB stacked bilayer graphene nanoribbon." *Current Applied Physics*, 2020, 20(4), 572.

1.6 Synopsis of book

Chapter 2 outlines the fundamental conception of carbon-based materials in order to explain the results in succeeding chapters. Molecular structure and orbital models of carbon nanomaterials have been discussed. This chapter also includes a brief discussion of computational technique and model that was used to achieve the results.

Chapter 3 explains the effect of combined isotope and vacancy on the phonon properties of BLG. At first phonon density of states is discussed and later mode pattern and localization length of the system are described.

Chapter 4 outlines the effect of edges and vacancies on the vibrational properties of bilayer graphene nanoribbon. At first, the effect of edges is studied which was later complemented by vacancy effects on the nanoribbons. The thermal conductivity of disordered nanoribbons is also described at the end of this chapter.

Finally, **chapter 5** contains a summary of this work. This chapter also points out the relevant possible future work.

Chapter 2. Theoretical Background

2.1 Introduction

This chapter describes some basic concepts of graphene. The fundamental vibrational properties of bilayer graphene are explained. The computational model used for this work is also described in this chapter.

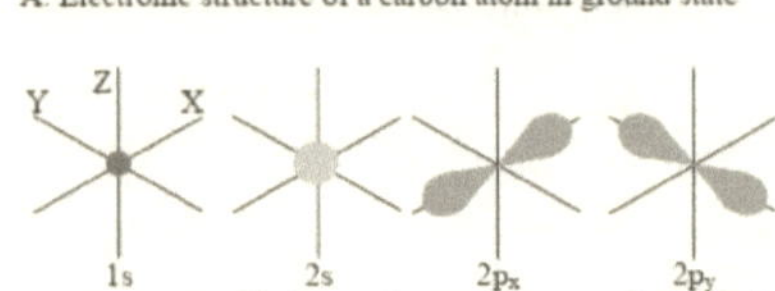

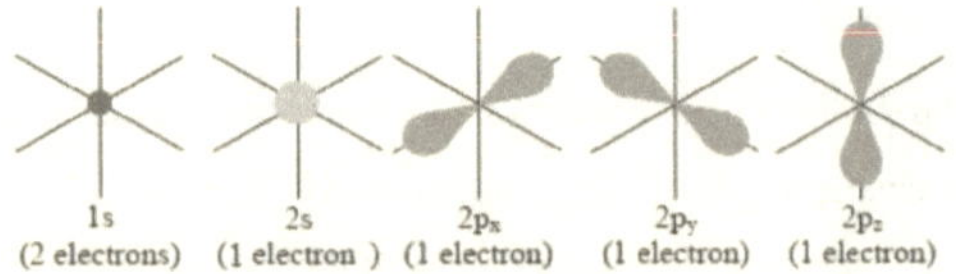

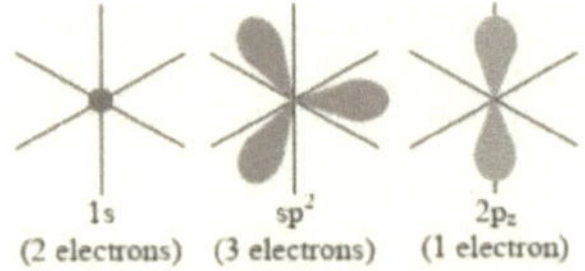

Fig. 2.1. Hybridization between carbon atoms to form sigma and pi bonds [56].

2.2 Fundamentals of sp^2 bonded carbon-based nanomaterials

Carbon is situated at group IV of periodic tables which is a very active material that is involved in many crystalline solids. Carbon atom consists of 6 electrons that are distributed as $1s^2$-$2s^2$-$2p^2$. Carbon atoms form any allotropes by sharing its four valence electrons. When carbon atoms take part in the formation of any crystal lattice, one electron from 2s orbital become excited by acquiring energy from the nuclei and jump to the $2p_z$ orbital. Thus, the net energy of the system decreases and this merging between 2s and 2p orbitals are commonly known as sp^2 hybridization and the newly formed orbitals are called hybrid orbitals. The formation of sigma and pi bonds through hybridization between carbon atoms have been illustrated in Fig. 2.1. The sp^2 hybridization between 2s orbital and two 2p orbitals containing only one electron is responsible for the sigma bonded trigonal planar structure between two neighboring carbon atoms. On the other hand, half-filled $2p_z$ orbitals form a pi bond by pairing up with other $2p_z$ orbitals from neighboring carbon atoms. This half-filled pi bond is mainly responsible for the semi-metallic properties of graphene. Fig. 2.2 shows the formation of a honeycomb lattice structure of SLG.

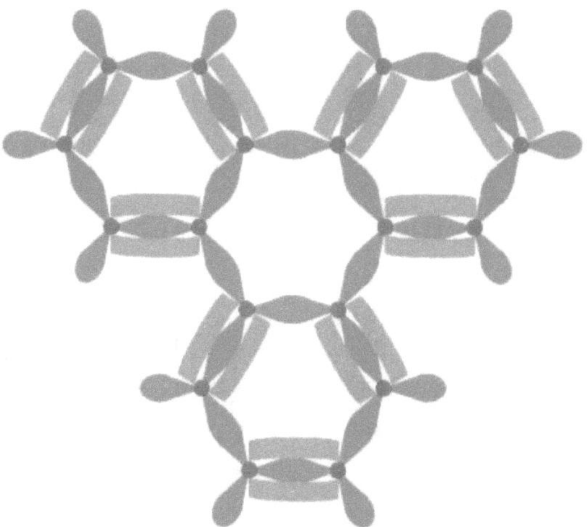

Fig. 2.2. Carbon atoms forming hexagonal structure in graphene lattice [56].

Charge carriers in graphene follows a linear energy dispersion relation, which is different from conventional electrons and holes conduction in other semiconducting materials. Thus, charge carriers in graphene behave like chiral massless particles, known as Dirac-fermions [57]. Therefore, conduction band minima and valance band maxima in graphene nearly overlaps at Fermi level forming a conical shape as shown in Fig. 2.3, where the Fermi-Dirac bandgap is zero.

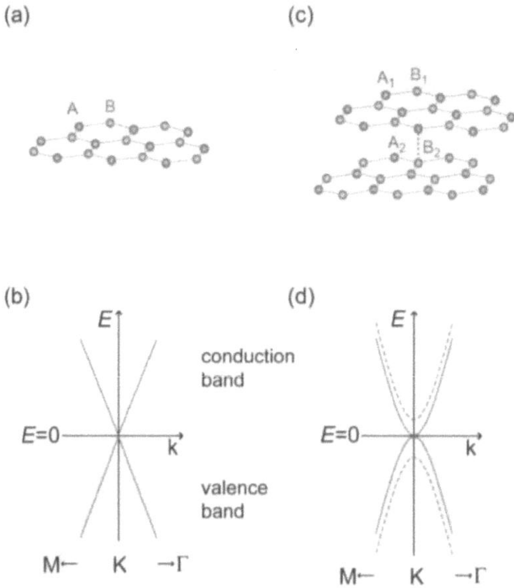

Fig. 2.3. On the left side (a) Titled view of SLG and (b) its schematic band structure. On the right side (c) titled view of BLG and (d) its schematic band structure. The dotted lines show the bandgap structure that could be achieved at the presence of vertical electric field in BLG [58].

Since charge carriers in pristine graphene are massless, that results in linear low energy band dispersion, move through the medium with energy independent Fermi velocity of around 10^6 m/s [58]. But at the presence of defects, energy bands become flattened at K-point that induces a small energy gap in graphene which further reduces the Fermi velocity of carriers.

2.3 Vibrational properties of graphene

From the dispersion relation, six phonon modes can be found for monolayer graphene which can be labeled as in-plane and out-of-plane, longitudinal and transverse, and acoustic and optical (iLO, iLA, iTA, iTO, oTO, oTA). But oTO and oTA are commonly referenced as ZO and ZA respectively. The phonon modes are labeled according to the direction of their movement along with the unit cell. Moreover, the phonon density of states (PDOS) contains all the information about associated phonons as shown in Fig. 2.4.

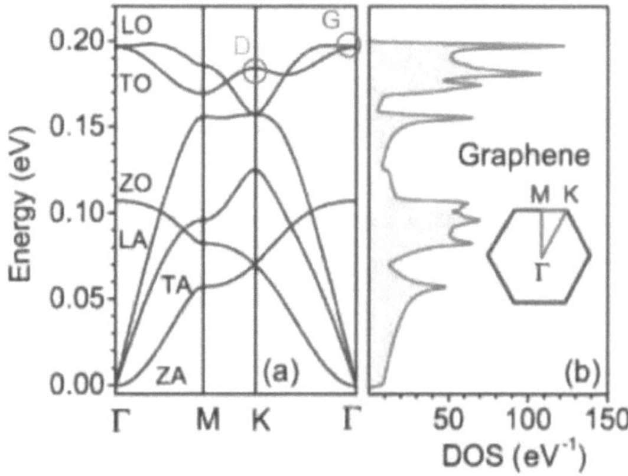

Fig. 2.4. Phonon dispersion and PDOS of SLG [59].

From Fig. 2.4 it can be noticed that K-point TO mode phonons are responsible for Raman active D peak whereas Γ-point LO-TO modes are responsible for the G peak of graphene. At Γ-point, degenerate LO-TO modes are found for graphene which is responsible for the E_{2g} mode phonon peak in the PDOS. The atomic motion of K and Γ point phonon modes have also been demonstrated as shown in Fig. 2.5 and Fig. 2.6. For a graphene sheet, the arrows from the circles denoting atoms point towards the direction of phonon displacements. Solid circles form the A sublattice while white circles denote B sublattice of graphene. The size of the circles in Fig. 2.6(c) and 2.6(f) represent the magnitudes of vectors which are $\sqrt{2}$ and $1/\sqrt{2}$ of the magnitudes of other vectors in Fig. 2.6(a)-(b) and (d)-(e).

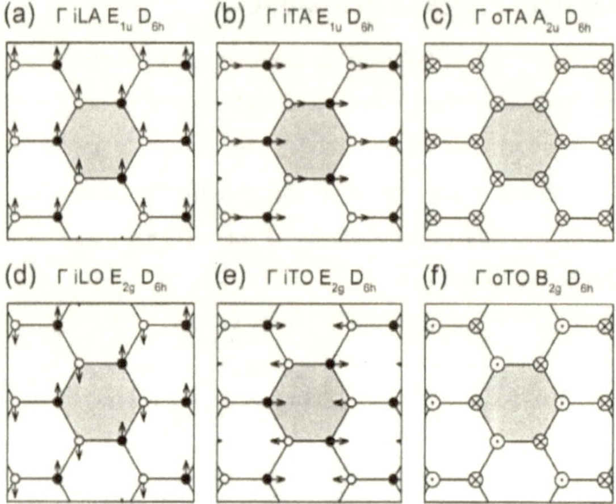

Fig. 2.5. Directions of different Γ point phonon modes of graphene. Cross and dot in (c) and (f) denotes the vector coming from out of and into the image plane respectively [60].

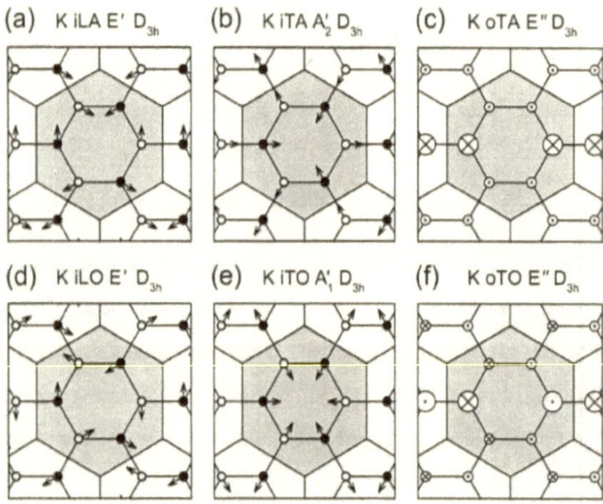

Fig. 2.6. Directions of different K point phonon modes of graphene. Cross and dot in (c) and (f) denotes the vector coming from out of and into the image plane respectively [60].

2.4 Raman spectroscopy of graphene

Raman spectroscopy is a technique to determine the number of layers, concentration percentage of defects, substitutional doping levels and much more. Raman spectrum of SLG has four major peaks denoted as D, G, D' and 2D. Fig. 2.7 demonstrates a typical Raman spectrum of graphene. The G peak is the result of degenerated LO-TO modes at the center of the Brillouin zone. Furthermore, K-point TO mode phonons are responsible for D peak which is achieved when a defect reaches a total momentum of zero as required for Raman scattering [27]. D peak position is dependent on the laser excitation energy since higher momentum transfer is needed at higher excitation energies. On the other hand, the position of the G peak is independent of excitation energy which is always at around 1590 cm^{-1}. For multilayer graphene, the 2D peak becomes broadened. For n (n>1) layer graphene (nLG), 2D peak encompasses 2D bands from nLG and (n-1)LG. Therefore, a broadened peak is achieved for nLG as demonstrated in Fig. 2.8.

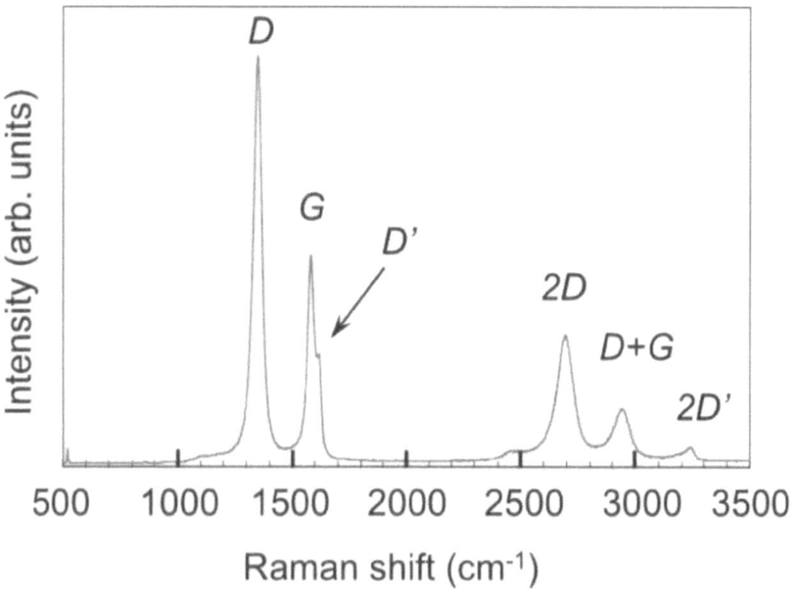

Fig. 2.7. Typical Raman spectrum of SLG [59].

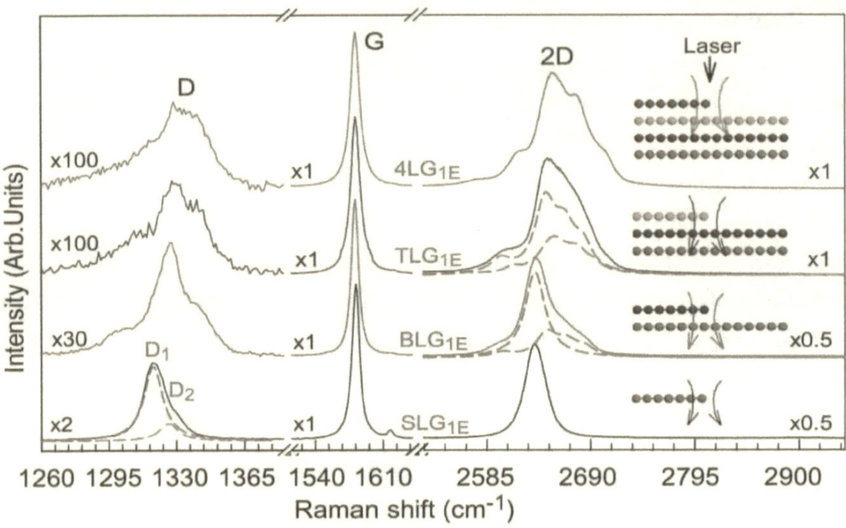

Fig. 2.8. Raman spectra of multilayer graphene showing high frequency regime [61].

2.5 Computational Method

2.5.1 Introduction to Force Vibrational (FV) method

FV method can be described for any linear atomic system such that when an external force is applied to that system with a known frequency, after some time all the atoms of that system will vibrate with large amplitudes at eigenmodes close to that frequency [55]. Therefore, this method can be applied to any disordered system very easily and can also be a tool for time-efficient calculation of vibrational properties of any system. The detailed technique for the calculation of PDOS has been explained in **Chapter 3** and **Chapter 4** in the computational modeling part for BLG sheet and nanoribbons respectively. Therefore, only the mode pattern calculation technique will be described here.

2.5.2 Mode pattern calculation

If any atomic system is driven at a frequency Ω for a sufficiently long time, the response will come from eigenmodes close to that frequency. Due to the localization effect, energy distribution becomes nonuniform. Furthermore, owing to the small number of excited modes, the chance of overlapping between modes is almost zero. Therefore, it can be assumed that only one mode is responsible for any high-density energy regime which is the base for the calculation of mode pattern of that system. Let's assume the displacements of lth atom at time t is-

$$u_l(t) = A_\lambda(t)\frac{d_l}{\sqrt{M_l}}$$

Equation 2.1

where A_λ is the amplitude of eigenmode λ, d_l is the displacement pattern or polarization vector at λ and M_l is the mass of lth atom. At $t = 0, u_l = 0$ and at $t = t$, if an external force $F_l cos\ (\Omega t)$ is applied to the lth atom, then the displacement will be-

$$u_l(t) = \frac{1}{\sqrt{M_l}}\Sigma_\lambda F_\lambda h(t, \omega_\lambda, \Omega)d_l(\lambda)$$

Equation 2.2

where,

$$h(t, \omega_\lambda, \Omega) = \frac{2}{\Omega^2 - \omega_\lambda^2}(\sin\ (^1/_2\ (\Omega - \omega_\lambda)t)\sin\ (^1/_2\ (\Omega + \omega_\lambda)t))$$

Equation 2.3

and,

$$F_\lambda = \Sigma_\lambda \frac{F_l}{\sqrt{M_l}}d_l(\lambda)$$

Equation 2.4

After that, eigenvectors of the system are calculated. When the whole model is driven for a time t', then the displacement of lth atom will be-

$$u_l^1(t) = \frac{1}{\sqrt{M_l}}\Sigma_\lambda F_\lambda h(t', \omega_\lambda, \Omega)d_l(\lambda)$$

Equation 2.5

Then the applied force F_l is modified to a new value-

$$F_l = u_l^1 M_l$$

Equation 2.6

After that, the whole system is again considered as at resting position and the system is again run for a time t' and this process goes on until rth iterations. Therefore, after rth iteration, the displacement of lth atom will be-

$$u_l^r(t) = \frac{1}{\sqrt{M_l}}\Sigma_\lambda F_\lambda h(t', \omega_\lambda, \Omega)d_l(\lambda)$$

Equation 2.7

After an ample amount of iterations, let's say eigenmode λ_1 gives the largest value for $h(t, \omega_\lambda, \Omega)$ function. Therefore, that mode is entirely depends on the frequency, not on the applied force F_l. In this way the spatial eigenmodes of any system can be quantified.

However, to apply FV method in BLG system, the first requirement is to calculate the position of the atoms and angle between atoms. According to the position of atoms, they can be divided as A type and B type atoms (Fig. 2.9(b)). The distance between intralayer A and B type atom is carbon-carbon bond length in graphene as denoted by a in Fig. 2.9(c). But distance between lower layer A type atom to upper layer B type atom can be calculated from Pythagoras formula as denoted by X in Fig. 2.9(c).

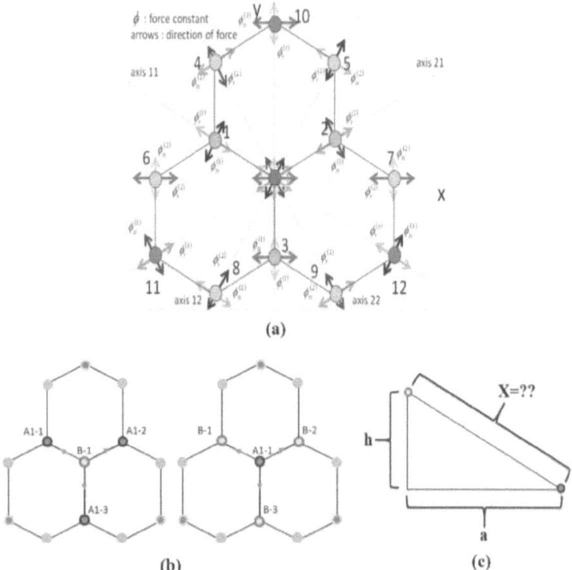

Fig. 2.9. (a) Schematic showing interaction direction between atoms in SLG for 3rd nearest neighbor from the central atom (red). [59]. (b) Top view of BLG by considering B type atoms as central atoms (on left) and A type atoms as central atoms (on right). (c) Distance calculation from lower layer A atom to upper layer B type atom.

Chapter 3. Interlayer isotope and vacancy effects on the phonon modes in AB stacked bilayer graphene

3.1 Introduction

The study on isotope enriched graphene samples [43, 62-65] has attracted enormous interest in the past few years. The growth mechanisms of graphene and carbon nanotube can be understood in a better way by the isotope labeling technique. The physical properties of graphene can be controlled in a greatly selective way using this process. Using the isotope labeling Li et al. [46] showed that graphene is grown by the CVD technique strongly depends on the substrate. Moreover, at high temperatures, adlayer graphene produced simultaneously between the first graphene layer and the Cu substrate. In addition, if we look into the spectral linewidth of Raman spectra of graphene, isotopes are considered to be responsible for intrinsic scattering mechanisms. By inducing isotopes in bilayer graphene (BLG), a thorough study on phonon properties can be done easily since isotope enriched samples exhibit modified phonon and thermal properties while maintaining the same electronic and chemical properties as in pristine sample [28]. Therefore, ^{13}C atoms in the system will not have any effect on the electron-electron interaction, which aids to distinguish easily the change in electron-phonon interaction. Higher mass atoms in a system disrupt the transmission of phonons, therefore phonons are responded strongly by the isotopic mass differences. These effects are perceived from spectroscopic tools like Raman spectroscopy, electron energy loss spectroscopy (EELS), infrared (IR) spectroscopy and so on.

Kalbac et. al. [63] studied specially fabricated BLG samples where an individual layer contains only ^{12}C (or ^{13}C) atoms to differentiate the Raman peaks responsible for individual layers. Moreover, by studying the linewidth of Raman G and 2D bands, Carvalho et. al. [64] found that phonon lifetime is dependent on the ^{13}C concentration in the graphene system. Costa et. al. [65] noticed a Raman shift and broadening of Raman G band with the increased concentration of ^{13}C atoms in single-wall carbon nanotube (SWCNT). In addition, they experimentally proved that isotope enrichment cannot alter the electronic properties of graphene.

On the other hand, the presence of a small number of vacancies in graphene system has the ability to change the low-temperature heat conduction properties. While there is no significant impact on electronic properties due to isotopes, vacancies have notable effects on electronic properties in the graphene system [31] which in turn influence the electron-phonon coupling. When defects are present, translation symmetry of the graphene system can be disrupted which in turn leads to the disintegration of momentum conservation through stimulating interior Brillouin zone K-point phonons. Consequently, the D band peak is observed in Raman spectra of disordered graphene sample, though it disappears for a pristine sample [32-34]. In any disordered system, phonon modes may arise outside of the acknowledged frequency regime of a pristine system [35-36], which can be termed as localized phonon modes alike Anderson's localization of electrons. Besides, optical mode phonons possessing shorter wavelengths are greatly affected by the isotopes and vacancies and become immobile. Chen et. al. [43] showed that the thermal conductivity of the isotope induced graphene system can be reduced approximately by 40%. This might occur due to the scattered phonons originated from mass mismatching in isotope disordered system. When phonons become localized, their mean free path tends to be finite and varied according to the square of the localization length. Since, phonons can be scattered at the presence of vacancies or ^{13}C isotopes in BLG, their lifetimes and mean free paths get confined into finite values. This might affect the Raman spectra of isotope and vacancy induced sample of BLG. Therefore, it is crucial to study the combined effect of isotopes and vacancies on phonons in BLG.

In recent years, some experimental works have been done on the phonon properties of bilayer graphene considering isotope effects [21, 66]. Lindsay et. al. [67] observed lower thermal conductivity in multilayer graphene (MLG) than SLG which is caused by interlayer interactions in MLG. They also showed that scattering caused by ^{13}C isotopes is responsible for much lower thermal conductivity in MLG. Rodriguez-Nieva et. al. [62] showed numerically that, phonon wave functions can be localized at the high energy optical mode phonon region owing to the scattered phonons from ^{13}C atoms in single-layer graphene (SLG). Employing molecular dynamics (MD) technique, Zhang et. al. [68] calculated the thermal conductivity of pristine graphene by taking out of plane phonons in consideration. They also claimed that a maximum of thousand-fold reduction in thermal conductivity of SLG can be possible with a high vacancy concentration. Furthermore, Van Tuan et. al. [69] analyzed the effect of dimer vacancies in BLG. They observed the ballistic regime with the increased dimer vacancies, while non-dimer vacancies are responsible for localized regimes.

Since there is lacking in the study of BLG by considering combined effect of isotopes and vacancies in, in my present work, I have extracted various phonon properties of isotope and vacancy disordered BLG. AB stacked BLG has been taken into account in my works owing to its notable sensitivity to band-gap tuning at the presence of external electric field and superior structural stability [70-71]. I have employed forced vibrational (FV) method [55] to my present model to compute the phonon modes. FV method mainly focuses on mechanical resonance to probe phonon eigenmodes from a sample of adequate size. I have considered hybrid layers with ^{12}C and ^{13}C atoms, while varied the number of isotopes from 0% to 100%. Furthermore, I have varied the concentration of vacancies up to a wider range in the bilayer graphene sample (with and without isotopes). In order to illustrate different phonon states, I have also computed the mode pattern of isotope and vacancy enriched bilayer graphene system. I lead my concentration mainly to the K-point in-plane transverse optical (iTO) mode phonons since they have a significant effect on Raman active D band peak. To briefly demonstrate the effect on phonon transmission in defect induced BLG, I have also computed the phonon localization length at the presence of various defect concentration. In addition, to illustrate the effect of different phonon modes on the Raman spectrum of isotope induced BLG, I have tried to approximate the Raman active G peaks from the computed phonon density of states (PDOS) employing convolution-deconvolution technique [72].

3.2 Theoretical Background

The implemented Force Vibrational (FV) method is based on the idea that an externally applied periodic force with a certain frequency to any system can assist the eigenmodes of that system to satisfy the resonance condition. Since the number of excited eigenmodes is directly linked with the energy of that system, it becomes possible to calculate the phonon density of states around the applied frequency as discussed below.

Considering a system containing N atoms where atoms are bound up with each other by linear springs, if a periodic force $F_l = F_0\sqrt{M_l}cos\,(\varphi_l)cos\,(\Omega t)$ with frequency Ω is applied from outside of the system, the equation of the oscillating motion of the lattice system can be defined as-

$$M_l\ddot{u}_l + \sum_{l'}\Phi_{ll'}u_{ll'}\,(t) = F_0\sqrt{M_l}\cos(\varphi_l)\cos(\Omega t) \qquad \text{Equation 3.1}$$

where $\varphi_{ll'}$ is the spring constant between the lth and l'th atoms, F_0 is a time independent constant, M_l is the mass of atom l and φ_l is an arbitrary value spreading over the range from 0 to 2π.

The eigenmode dependent displacements can add up to as follows-

$$u_l(t) = \sum_\lambda A_\lambda(t)\frac{e_l(\lambda)}{\sqrt{M_l}} \qquad \text{Equation 3.2}$$

where λ denotes the eigenmodes and A_λ (t) and e_l are time dependent amplitude and eigenvector of λ respectively. Consequently, over a long time the eigenmodes get excited around the applied frequency, Ω. By considering the orthogonality of eigenvectors and taking mean φ_l, the mean energy can be written as-

$$< E(\Omega) > \approx \frac{\pi t F_0{}^2}{8}\sum_\gamma \delta(\omega_\gamma - \Omega) = \frac{\pi t F_0{}^2 N g(\Omega)}{8} \qquad \text{Equation 3.3}$$

Therefore, the phonon density of states $g(\Omega)$ is-

$$g(\Omega) = \frac{8 < E(\Omega) >}{\pi t F_0^2 N} \qquad \text{Equation 3.4}$$

At the presence of external force over a long time, if the modes around applied frequency Ω get localized, there will be consistent energy distribution all over the system. The probability of overlapping excited modes is very little owing to the smaller number of excited modes at Ω. This phenomenon leads to the fact that only a particular mode is accountable for any region with an intense amount of energy. Accordingly, the displacement of the enclosed atoms into a localized region can be written as-

$$u_{c,l} = A\frac{e_l(\lambda_0)}{\sqrt{M_l}} \qquad \text{Equation 3.5}$$

where A is a uniform quantity for all l and $e_l(\lambda_0)$ is the polarization vector at mode λ_0. In this way, we can find the mode pattern of the lattice system.

Localization length can be quantified by computing the inverse participation ratio, IPR_λ which is determined by the second-order moment of the displacement at mode λ as follows [73]-

$$IPR_\lambda = \frac{\sum_{l=1}^N |u_{l,\lambda}|^4}{(\sum_{l=1}^N |u_{l,\lambda}|^2)^2} \qquad \text{Equation 3.6}$$

Where $u_{l,\lambda}$ is the displacement of the atom at lth position at eigenmode λ. Since the vibration of atoms in the localized region is very limited, it can be assumed that m number of atoms is in oscillating motion. Therefore, taking orthogonality of eigenvectors into account, the amplitude of oscillation can be written as $a = 1/\sqrt{m}$. In addition, for extended phonon mode, $IPR_\lambda = 1/N$ which in turn suggests that IPR_λ varies in the range from 1/N to 1 for any general phonon mode. Furthermore, IPR_λ helps to define the localization length ζ_λ, since $\zeta_\lambda \propto IPR_\lambda^{-1/2}$ [74]. At 0% defect concentration in any lattice system, all the phonon modes become extended which in turn leads to the following equation-

$$\frac{\zeta_\lambda}{\zeta_0} = \left(\frac{IPR_0}{IPR_\lambda}\right)^{1/2} \hspace{3cm} \text{Equation 3.7}$$

where IPR_0 is the mean value of IPR_λ for a lattice system with 0% defect concentration and ζ_0 is the size of that system. Thus, the localization length of any system at mode λ can be calculated.

3.3 Computational Model

FV method is employed to compute the phonon properties of the pristine bilayer graphene sheet as well as isotope and vacancy induced sample by considering the free boundary conditions (FBCs). To illustrate, the far-left atom will only be restored by its right atoms every time it oscillates from its equilibrium state. Furthermore, force constants for the atoms outlying from the graphene edge will be more like those under periodic boundary conditions (PBCs). In my observation, I ignore the effects of strains, surface reconstruction and torsions to keep it simple and approximately 10600 atoms have been considered to form the pristine bilayer graphene honeycomb structure, i.e. 5300 carbon atoms in each layer. Employing site percolation theory, vacancies and ^{13}C isotopes have been induced arbitrarily to the system. Vacancy concentration was varied over a long-range from 0 to 30%. Previously, one study showed that the chemical reduction of graphene oxide can produce a graphene sample comprising almost 9% of vacancy [75]. While another study proclaims that, vacancy concentration up to 30% is attainable in graphene oxide [76]. Moreover, molecular dynamics simulation elucidated that graphene can retain its honeycomb structure from irradiation-induced damage for vacancy concentration of up to 35% [77]. Thus, in my model variation of vacancy concentration has been considered up to 30%. Force constant tensors up to 4th nearest neighbor were utilized in the model for both interlayer and intralayer interactions, which have been tabulated in Table-3.1 and Table-3.2 respectively.

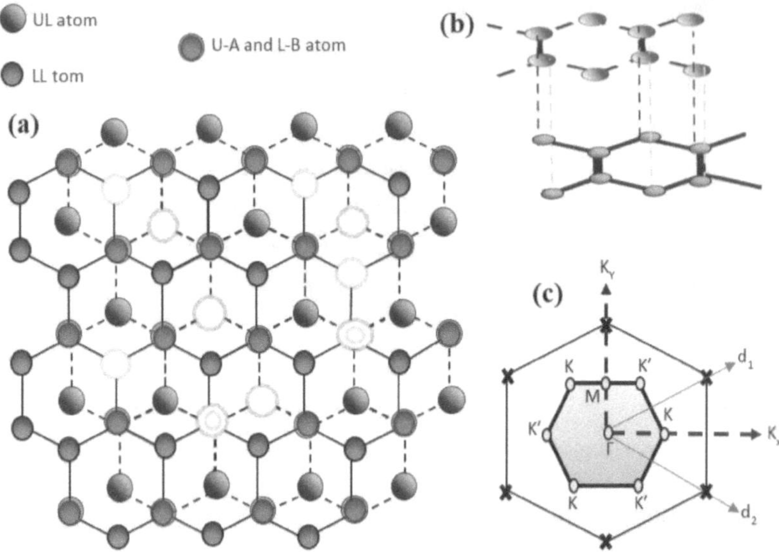

Fig. 3.1 (a) Top view of A-B stacked bilayer graphene. Upper layer A and B type atoms are represented by U-A and U-B while L-A and L-B denote lower layer A and B type atoms respectively. The shadowed red and black circle shows the randomly induced vacancy sites. (b) Side view of A-B stacked bilayer graphene lattice. (c) Reciprocal lattice of graphene (single layer), where crosses represent the lattice points and d_1 and d_2 are reciprocal lattice vectors. The shadowed hexagon is the first Brillouin zone of graphene with high symmetry points Γ, K, M.

Owing to the higher mass of 13C isotope atoms, the dynamics of the lattice structure will be modified. Hence, in my computation isotopes have been varied over a range of 0 to 100%. In addition, the combined effect of both isotopes and vacancies have been taken into account in my simulation to have a better understanding. For atoms near edges which are only under the force constant of FBCs have been considered instead of PBCs.

Table-3.1: The out-of-plane force constants used in my computation for interlayer interaction of BLG as found by Jishi et. al. for graphite [78]. The units are N/m.

Neighbor	1st	2nd	3rd	4th
φ_{ra}	2.7469	0.59552	-0.47388	0.2
φ_{in}	-5.9341	1.2712	0.4739	-0.9278

Table-3.2: The force constants used in my computation for intralayer interaction of BLG as found by Jishi et. al. [79]. The units are N/m.

Radial	Tangential in-plane	Tangential out-of-plane
$\varphi_{ra}(1)=365.0$	$\varphi_{in}(1)=245.0$	$\varphi_{out}(1)=98.2$
$\varphi_{ra}(2)=88.0$	$\varphi_{in}(2)=-32.3$	$\varphi_{out}(2)=-4.0$
$\varphi_{ra}(3)=30.0$	$\varphi_{in}(3)=-52.5$	$\varphi_{out}(3)=1.5$
$\varphi_{ra}(4)=-19.2$	$\varphi_{in}(4)=22.9$	$\varphi_{out}(4)=-5.8$

3.4 Results and Discussion

Six phonon branches are found in any pristine single layer graphene; 6n branches for n layer graphene. A low frequency regime peak is observed at ~100 cm^{-1} for bilayer graphene which is relevant to Γ-point out of plane ZA_2 mode phonons in bilayer graphene. In addition, doubly degenerate Γ-point longitudinal optical (LO) and transverse optical (TO) mode phonons are accountable for the origin of G peak near 1590 cm^{-1} owing to the scattering between them. Fig. 3.2 depicts the computed PDOS of bilayer graphene with 0 to 100% isotopes in both layers (i.e. $(^{13}C+{}^{12}C)/(^{12}C+{}^{13}C)$ hybrid structure) as well as only in the upper layer (i.e., $^{13}C/^{12}C$ structure). For pristine bilayer graphene all corresponding phonon peaks are found, which is very similar to previous study by Jishi. et. al. on graphene-based system where force constant method (FCM) was employed [79]. The simulated PDOS also unveils the fact that, with the increase of ^{13}C atoms in the system, LO and TO phonons become nondegenerate at the center of BZ which in turn give rise to the distorted peaks in the PDOS. The shift of origin frequencies associated with Raman G and D band peaks for isotope enriched graphene system can be approximated as [65]-

$$\omega_{12_{(1-X)}{}^{13}C_X} = \omega_{12_C}\sqrt{\frac{12}{12(1-X)+13X}} \qquad \text{Equation 3.8}$$

where X is the isotope concentration in fractional format over the range of 0 to 1 and ω_{12_C} denotes the phonon mode frequency for pristine graphene system. Therefore, this equation suggests vibrational frequency to be inversely proportional to the square root of the atomic mass of the system. The computed PDOS figures have also justified the equation since a downshift of around 70 cm^{-1} for E_{2g} mode has been observed upon the 100% ^{13}C atoms enrichment in both layers of BLG lattice as shown by the dashed line in Fig. 3.2(a). Though for SLG containing 100% ^{13}C atoms, a downshift of nearly 60 cm^{-1} was found in previous study [80]. This concludes the effect of interlayer Van der Waals attraction on the phonon-phonon scattering along the out of plane direction which leads to the change in phonon modes of BLG. Moreover, when the isotope percentage is not perfectly 0% or 100%, but actually in between them, then translation symmetry breaks down and mass mismatch takes place. Thus unfortunately, the wave vectors of corresponding unit cell hardly comprise good quantum numbers. Therefore, phonons become scattered in different states which leads to the broadened E_{2g} mode peaks as seen from Fig. 3.2 for 25% to 75% isotope concentration. Moreover, low frequency regime Γ-point ZA_2 peak has been observed in Fig. 3.2 for BLG, which is absent in SLG. In addition, I have plotted the gradual downward shifting of E_{2g} mode peak frequencies extracted from Fig. 3.2(a) which has been shown in Fig. 3.3. I have also plotted the calculated E_{2g} mode frequencies from above equation as marked by the solid lines in Fig. 3.3. Furthermore, some strange peaks can be observed at the higher frequency regime for the PDOS of the sample where only the upper layer contains 100% isotopes as illustrated in Fig. 3.2(b). To narrate those peaks, I tried to estimate the Raman spectrum of bilayer graphene by employing convolution-deconvolution method introduced by Wang et. al. [72]. This method can act as a tool to approximate the Raman active peaks from PDOS of any system and this was described in details for graphene-based system in previous paper [81]. The estimated Raman G band peaks over the range of 1300 to 1700 cm^{-1} for both type of samples have been shown in Fig. 3.4. We can notice band splitting properties for the sample where upper (lower) layer contains only ^{13}C (^{12}C) atoms as indicated by dashed lines in comparison with pristine samples (either of ^{12}C or ^{13}C atoms). Since there is no mass mismatch in particular layers of that sample, two unique G band split peaks have appeared, which is very similar with the findings by Kalbac et. al. [63]. Raman active G peak was shifted downward by around 70 cm^{-1} for the BLG containing 100% ^{13}C atoms in both layers. In addition, no D band peaks were observed for any of the samples mentioned in Fig. 3.4 which suggests a good symmetry within individual layers of bilayer system.

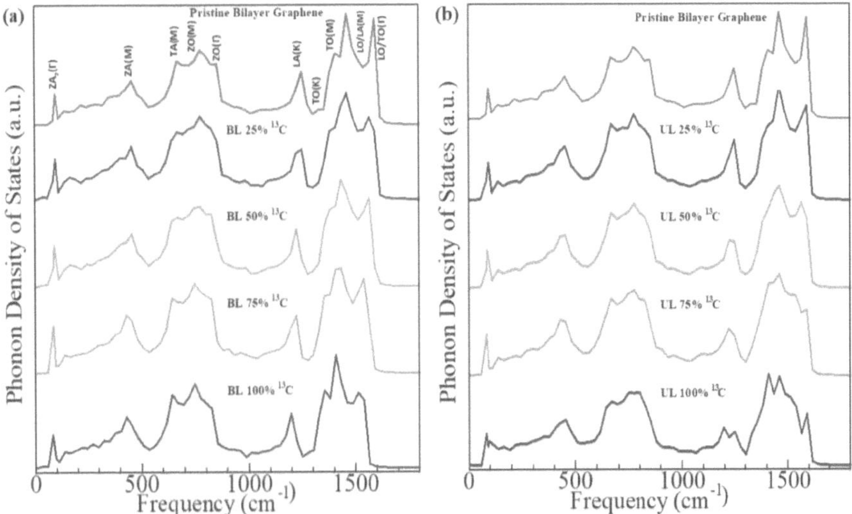

Fig. 3.2- Isotope induced PDOS of bilayer graphene for (a) both layer and (b) upper layer. BL and UL denote both layer and upper layer respectively. All major phonon peaks have been defined for pristine bilayer graphene in figure (a).

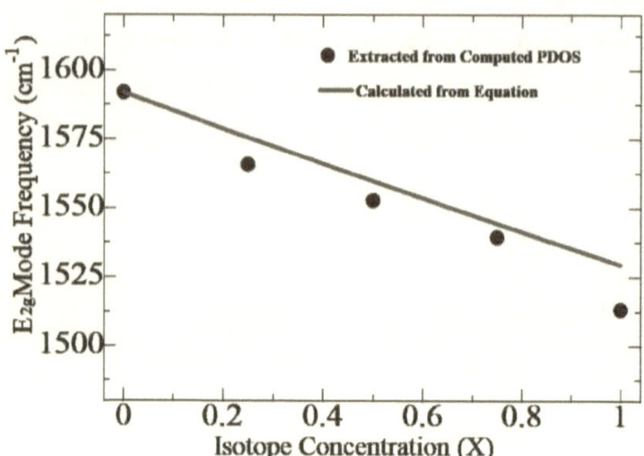

Fig. 3.3. Shifting characteristics of E_{2g} mode frequencies with different concentrations of ^{13}C atoms in both layers of bilayer lattice.

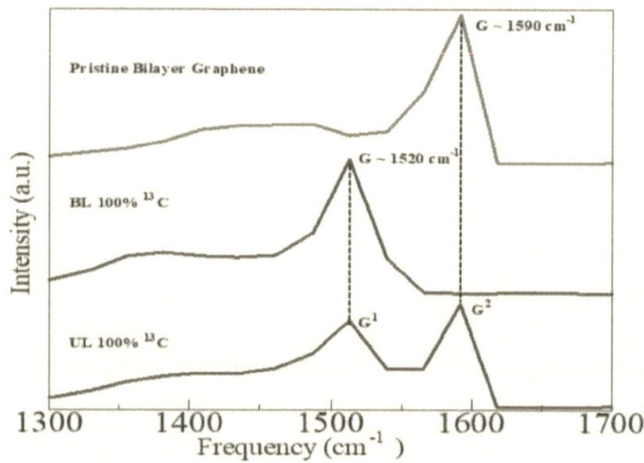

Fig. 3.4. Estimated Raman G band peaks for different isotope enriched samples. When only upper layer is induced by 100% isotopes, G peak splits into two which are represented by G^1 and G^2.

While I studied the PDOS of vacancy induced samples I observed flattened E_{2g} mode peak with the increase of vacancies as displayed in Fig. 3.5. A more prominent softening of the E_{2g} mode peak can be noticed for the samples where vacancies were induced in both layers other than only in the upper layer. When the vacancy concentration reaches up to 30% in both layers, the E_{2g} mode peak completely disappears as noticed in Fig. 3.5(a). Thus, the long-range order of carbon atoms in graphene can be collapsed by the vacancies when it reaches beyond a particular extent. Furthermore, some well-defined peaks can be observed at the low-frequency region which suggests the higher density of states of phonon modes at a few low-frequency regimes. The unsaturated dangling bonds of carbon atoms at high vacancy concentration give rise to this phenomenon and are also responsible for the reduced density of phonon modes at higher frequency regime.

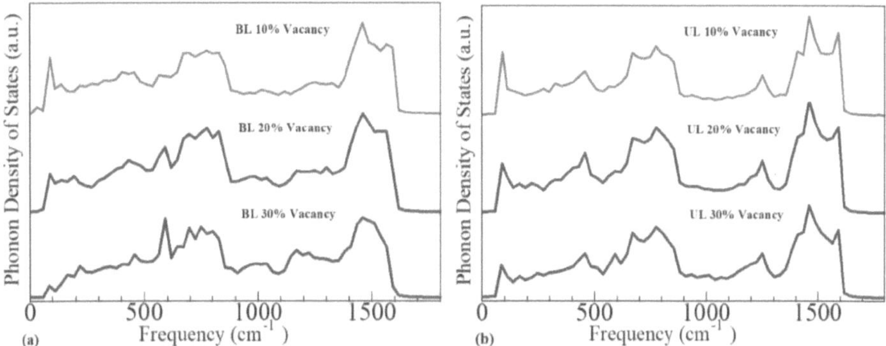

Fig. 3.5- PDOS for different concentrations of vacancies in (a) both layers and (b) upper layer of the bilayer graphene lattice. BL and UL represent both layer and upper layer respectively.

In addition, when I focus my concentration on the samples containing both vacancies and isotopes, a notable amount of broadening and downshifting can be observed for the E_{2g} mode peak as demonstrated by Fig. 3.6. For samples comprising isotopes as well as vacancies in both layers, demolishing of the E_{2g} peak becomes worse as illustrated in Fig. 3.6(a) and this is also explainable by the facts mentioned previously for vacancy induced samples. Although some new peaks can be observed in the low-frequency region, the average PDOS for both of the samples mentioned above is less than the pristine lattice. It could be caused by the inelastic behavior of flexural phonons in defect induced samples. Some previous studies showed that acoustic flexural phonons can demonstrate quadratic dispersion relation in independent pristine graphene suggesting a higher amount of low energy phonon modes [82-83]. On the other hand, flexural dispersion becomes more inelastic owing to the tension induced by defects or contacts which acts as a catalyst for an overall reduction in PDOS.

27

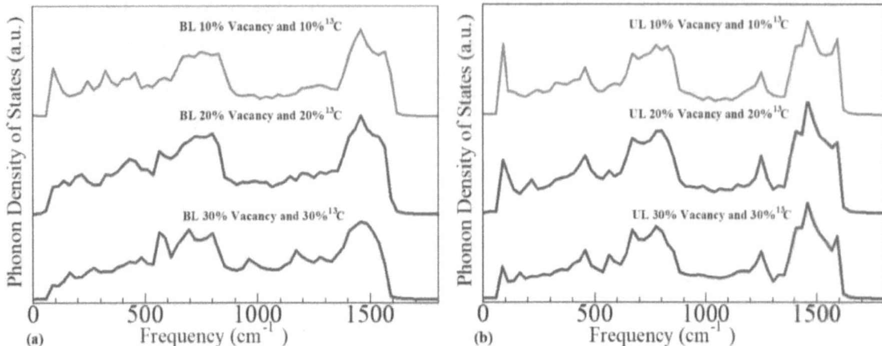

Fig. 3.6- PDOS for combined isotopes and vacancies induced in (a) both layers and (b) upper layer of bilayer graphene.

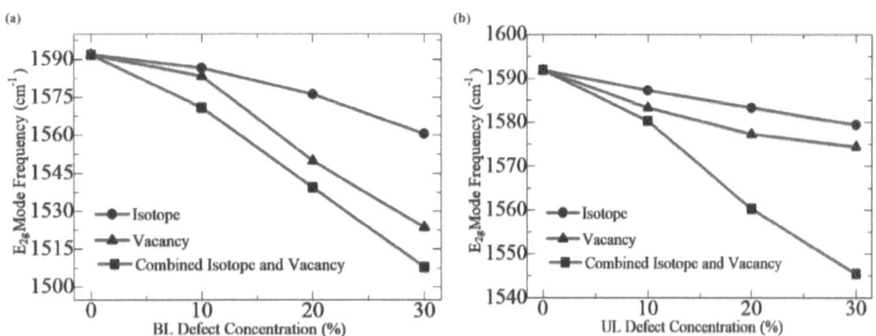

Fig. 3.7. Variation of E_{2g} mode phonon frequencies with respect to the various defect concentrations in (a) both layers and (b) upper layer of bilayer graphene system.

Furthermore, I have also compared the downward shifting characteristics of E_{2g} mode peaks for all types of samples up to a defect concentration of 30% as sketched in Fig. 3.7. Significant shifts can be observed with the rise in defect concentrations. Therefore, there will be a crucial modification in the electron transport properties of bilayer graphene with the increase of defect concentration, mostly for combined isotope and vacancy defects.

28

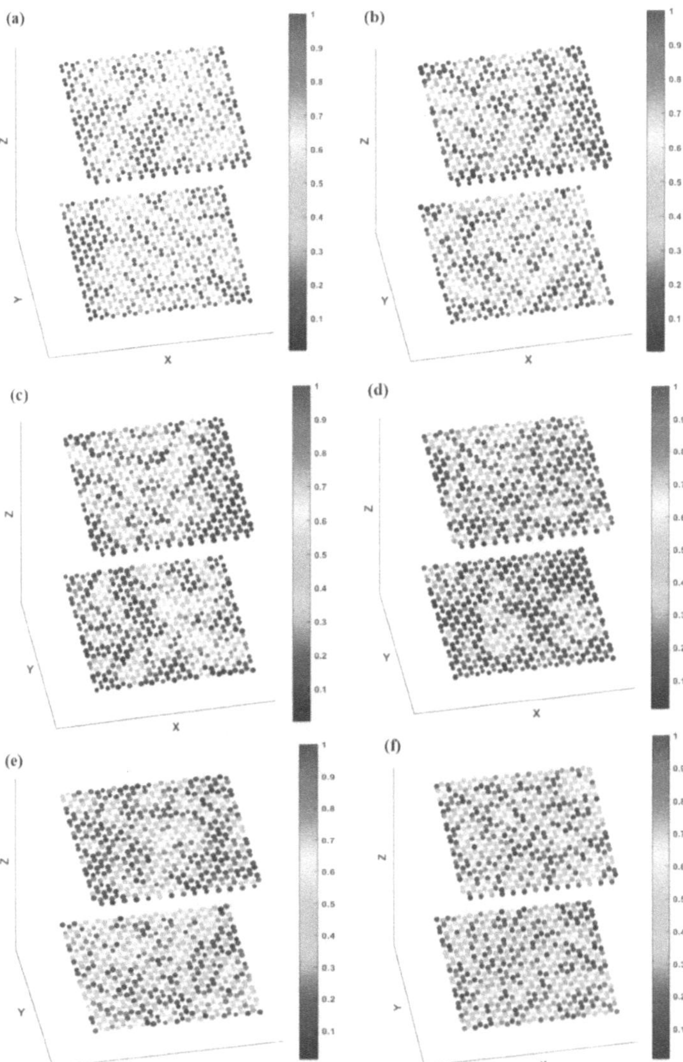

Fig. 3.8. Typical mode patterns for E_{2g} mode phonon of (a) pristine bilayer graphene and ^{13}C isotope induced sample with (b) 20%, (c) 40%, (d) 60%, (e) 80% and (f) 100% densities at $\omega \approx 1590$ cm^{-1}.

Any type of defects hinders the phonon transmission into a limited region. Vacancy and isotopes introduce disintegration in the symmetry of the lattice system. Thus, as outlined previously, wave vectors lose their good quantum numbers and phonons dispersed into different states. Therefore, there is a possibility for the phonons to become localized in real space. To visualize the phonon localization phenomenon, I have calculated typical mode patterns from the eigenstates of the atomic vibration. Typical mode patterns of pristine BLG as well as $(^{13}C+^{12}C)/(^{12}C+^{13}C)$ BLG structures have been sketched in Fig. 8. Here, I concentrate mainly on the in-plane longitudinal and transverse optical (iLO and iTO) phonons at $\omega = 1590$ cm^{-1} owing to the main contribution of these phonons on the G band in the Raman spectra. I have taken approximately 1100 atoms in consideration to display the mode pattern since the phonons become strongly localized at a frequency near 1590 cm^{-1} within this lattice configuration. The change of colors in Fig. 8 represents the displacement of the atoms.

As indicated in Fig. 3.8(a), modes are extended through the whole lattice points for pristine BLG sample. However, those extended modes mitigate when impurities are added to the system. As shown in Figs. 3.8(b)-(f), eigenmodes are spatially localized in a particular region for isotope induced samples which are caused by the pulsating oscillation from ^{12}C atoms coupled with ^{13}C isotopes. I differentiated the isotope atoms by denoting it as a bigger solid circle. As the Figs. 3.8(b)-(e) suggest, modes are strongly localized with increasing isotope impurity densities. However, among the different structures the strongest localization is observed at 60% ^{13}C isotope concentrations. This asymmetric behavior will be explained in the later section.

In addition, I have also displayed the mode patterns for the samples containing only vacancies as well as combined isotope and vacancies in both layers of BLG as illustrated in Fig. 3.9(a) and Fig. 3.9(b) respectively. When vacancies or other substitutional defects present in the graphene related system, a characteristic peak called D-peak along with the G-peak is observed in the Raman spectra. In-plane optical mode phonons at K-point play the vital role for the generation of this D band at ~1350 cm^{-1} in the Raman spectra. Hence, for vacancy or merging ^{13}C isotope and vacancy defected BLG sample, I concentrate my focus on the iTO phonon at 1350 cm^{-1}. If I analyze thoroughly, only a few highly displaced or dispersed phonon eigenmodes can be noticed around the defect sites compared to the pristine BLG sample. This indicates that K-point iTO phonon at 1350 cm^{-1} is also strongly localized near the vacancies or a combined defect sites which is conceptually similar to the D-peak in the Raman spectra of defected graphene systems. Since a random external force was employed to extract the phonon eigenmodes, the centers of localized modes are found in different positions with various time developments. However, in all cases centers of the localized modes are found near the defect sites.

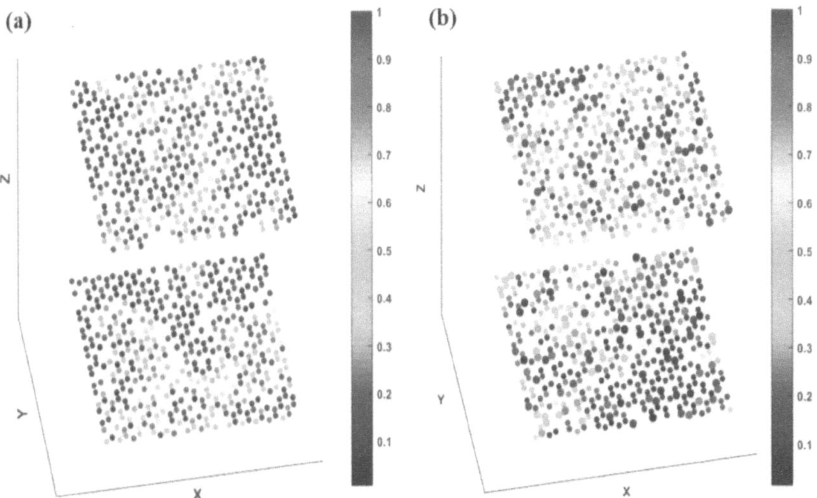

FIG. 3.9. Typical mode patterns for K-point iTO mode phonon of (a) 20% vacancy and (b) 20% combined isotope and vacancy induced bilayer graphene at $\omega \approx 1350$ cm^{-1}.

Next, I show the change in average localization length at E_{2g} mode frequency (~1590 cm^{-1}) with the increase of isotope concentrations from 0% to 100% in hybrid (^{13}C+^{12}C)/(^{12}C+^{13}C) and ^{13}C/^{12}C structures as illustrated in Fig. 3.10(a) and 3.10(b), respectively. The solid circle denotes the mean value as I ran the simulation ten times to minimize the fluctuations with different time development. As described in the "theoretical background" section, IPR_λ was computed to approximate the phonon localization length. If all the modes are extended then $IPR_\lambda = 1/N$, while $IPR_\lambda = 1$ when all modes become localized. Since in my simulation, an enormous system i.e. N=10680 atoms were considered, most of the states will be extended.

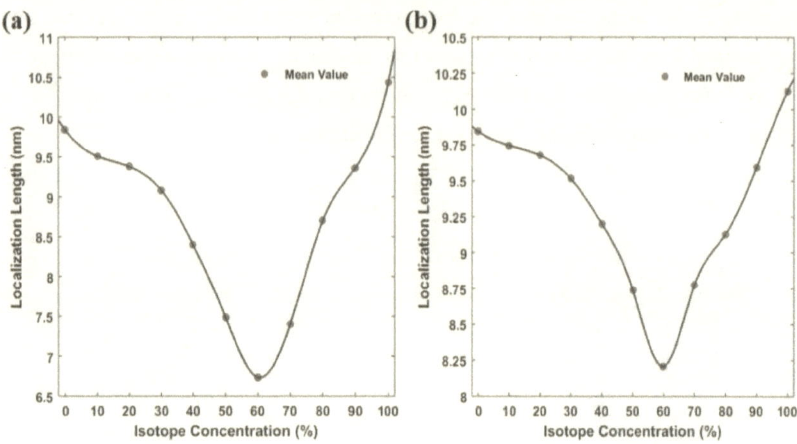

Fig. 3.10. Localization length of bilayer graphene at the E_{2g} mode frequency with various isotope concentration in (a) both layers and (b) only in the upper layer of the lattice.

For both types of samples, the localization length decreases abruptly at first up to isotope concentrations of 60% and then it shows an upward trend. Although an asymmetric behavior is found, however, these results are well consistent with the localization length of single-layer graphene by Rodriguez-Nieva et al. [62]. Many factors may induce this asymmetric nature. Increasing the mass of a few number of atoms in the lattice (inserting ^{13}C atoms to a ^{12}C lattice) does not have the same impact as reducing the mass of a few number of atoms in the lattice (inserting ^{12}C atoms to a ^{13}C lattice). The localization length is estimated from the IPR_λ since $\zeta_\lambda \propto IPR_\lambda^{-1/2}$, and IPR_λ is evaluated from the eigenstates of the atomic vibration. When I increase the mass of a few numbers of atoms in the system, the vibrational amplitudes related to those atoms become shorter than the overall amplitude of the eigenmodes of pristine lattice and vice versa. Moreover, the vibrational frequency is inversely proportional to the square root of the atomic mass (i.e. mass M as $M^{-1/2}$) of the system which makes significant changes in the PDOS with changing ^{13}C isotope concentrations. Therefore, the comparison of localization lengths at different isotope concentrations is not straightforward when measuring over a frequency window.

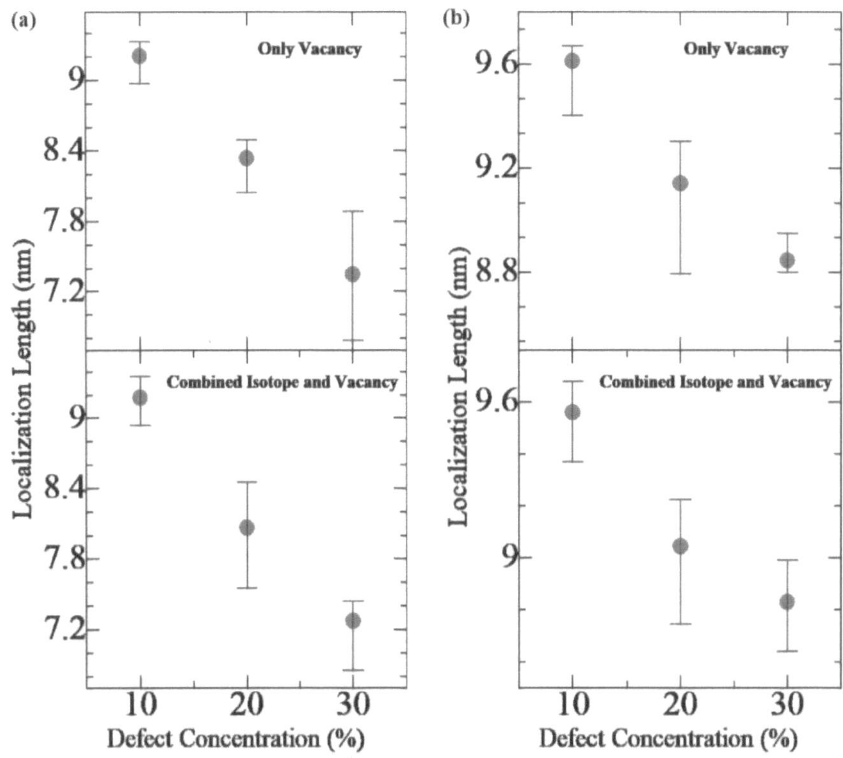

Fig. 3.11. Localization length of bilayer graphene at $\omega \approx 1350$ cm^{-1} with different defect concentrations in (a) both layers and (b) only in the upper layer of the lattice.

Finally, the localization length for vacancy as well as combined isotope and vacancy induced samples has been illustrated in Fig. 3.11 at $\omega \approx 1350$ cm^{-1}. With the increase of defect concentrations, localization length shows a downward trend which indicates higher localization states. For combined isotope and vacancy induced samples, the localization length reaches to a knee-high point in comparison with only isotope or only vacancy induced samples. The drop in localization length is more notable for samples where defects were induced in both layers (Fig. 3.11(a)) as compared to only in the upper layer (Fig. 3.11(b)). The resonant vibration between ^{12}C main atoms and defect sites are responsible for the downtick in localization length with the increase

33

of defect concentration. However, random energies at K point optical mode phonon might not only be responsible for observed phonon localization. It is feasible that defect induced backscattering at higher frequency regime can also be accountable for localized states. The computation was done at iTo mode frequency which justifies that region and can be compared to previous studies for monolayer graphene [37, 62]. The vibrational frequencies around the defect sites are different from the rest of the lattice which in turn gives rise to localized eigenstates.

3.5 Conclusion

To summarize, combined ^{13}C isotope and vacancy effect on the phonon properties of bilayer graphene has been studied. Phonon density of states has been computed for samples where ^{13}C isotopes have been varied over an ample range possible and vacancies were induced as rational as possible to a practical scenario. From the simulated results, it can be observed that PDOS of bilayer graphene is greatly impacted by the defect concentration and with the increase of defect concentration more broadened PDOS spectrum has been observed. Moreover, defect induced lattice vibrations are found to be responsible for the high peaks in the low-frequency regime of disorder bilayer graphene. Furthermore, a split of Raman active G peak has also been noticed for samples where ^{13}C isotopes have been induced in one layer of the bilayer sample. In addition, the combined effect of isotopes and vacancies on the localization of optical phonons have been represented. Mode pattern and localization length have been quantified for K-point in-plane transverse optical mode phonons. Localized modes have been noticed around the defect sites and localization length has been found to its lowest for combined isotope and vacancy disordered samples. Finally, localization length was computed for E_{2g} mode frequency at ~1590 cm^{-1} for a wide range of isotope concentration. At first, a decrease and then an increase in average localization lengths have been observed with the increase of isotope concentration.

Chapter 4. Impact of vacancy defects on the phonon modes in AB stacked bilayer graphene nanoribbon

4.1 Introduction

Phonon properties in graphene-related materials are very important phenomena to study since strong electron-phonon coupling exists in graphene-based system [84-87]. Moreover, heat dissipation is an important challenge in nanoelectronic devices, which strongly depends on the phonon modes of the semiconducting materials. In recent years, phonon properties of one dimensional (1D) nanostructured graphene nanoribbons (GNRs) have drawn much attraction since they show intriguing features that are distinct from two dimensional (2D) or bulk materials. In particular, bilayer graphene nanoribbons (BLGNRs) are of special interest as the width and interlayer interactions affect the phonon modes, hence show the fascinating electronic and thermal properties. It is demonstrated that the thermal conductivity in the 1D GNR channel increases [75], and follows the power-law function dependence, whereas the thermal conductivity in the 2D graphene channel shows the logarithmic trend [86].

In GNRs the edges are treated as defects and show intriguing electron and phonon properties because of its lower co-ordination number [88]. During the fabrication of GNRs using various techniques like Electron-beam lithography, Scanning Tunneling Microscopy, Electro-hydrodynamic nanowire lithography [29], it is also obvious that vacancy and ad-atom vacancy defects may exist inherently in the systems. In addition, stable interlayer bonds can take place in a multilayer graphene system due to the interaction between lower coordinated atoms and neighboring layers' vacancies [30]. Thereby, a more complicated structure combining the vacancies and edges is formed in BLGNRs. All of these defects might change the vibrational properties of BLGNR, thus play an important role in transport properties. A detailed understanding of these merging effects on the phonon modes is thus indispensable to know the transport properties in BLGNRs. When defects are present in any graphene lattice system, phonon modes are supposed to exist beyond the accepted frequency regime of the pristine system. This phenomenon is known as phonon localization, which is analogous to Anderson localization that depicts the localized states of electrons in any defective lattice system. Although localized edge phonon modes are reported for the graphene nanoribbon system [89], very little is known about the combined effects on the phonon modes of BLGNRs.

Formerly, Loh. et al. showed that acoustic modes of localized sites are clustered at the edges and the vicinity of the vacancies in GNRs [90]. Their spatial analyses revealed that thermal conductivity decreases with an increasing amount of vacancy disorder, where localized acoustic modes are the leading heat carriers. It is found in the literature that governing the amount of vacancy in any GNR system, the extended states can be isolated from the localized states [91]. Savic et al. exposed that, due to wave interference responses, randomly introduced vacancy disordered nanoribbon system exhibits three unique transport regimes, namely diffusive, ballistic and localized [37]. However, as far as my knowledge goes, no exploratory work has been done of the localization effects on the phonon wave function of the vacancy disordered BLGNR systems. Previously, Xu et al. pointed out the edge effect on the localization regimes of BLGNR [92]. In my prior work [93], I investigated the localized phonon modes in vacancy induced monolayer GNR system, where I revealed that localized modes are dependent on the vacancy concentration. Though recent studies explore the phonon localization, thermal conductivity, vibrational modes of monolayer and bilayer graphene nanoribbons [88, 90, 93-103], analysis on phonon modes of vacancy disordered BLGNR remains scarce.

Many computational techniques have been introduced earlier comprising density functional theory [104-105], force constant fittings [106-107], first-principles calculations [88, 100], to study the vibrational properties of the nanoribbon system. Lattice system becomes complicated in the presence of defects; thus, the dynamical matrix technique requires huge time for computing as well as convergence problem, which limits the system size up to a few atoms. Nonetheless, a sufficiently larger system is required to investigate the defect induced phonon properties of BLGNR due to its complex structure, hence, a reliable model should be developed to study the phonon modes of the disordered nanoribbon system.

In the present work, I have theoretically probed the vibrational properties of vacancy disordered AB stacked bilayer armchair graphene nanoribbons (BiAGNR). As vacancy concentration has a great effect on the phonon properties of BiAGNR, I alter the defect density up to a broad range. Moreover, to illustrate the effect of localized vibrational properties on disordered BiAGNR, I have computed the mode pattern for pure and defective BiAGNRs. I have shown absolute curiosity on the K-point in-plane transverse optical (TO) modes due to its important inference in the D band of Raman spectra. Furthermore, I have also analyzed the impact of vacancy induced phonon modes on the heat conduction properties such as specific heat capacity and thermal conductivity in the BiAGNR system.

4.2 Computational Model

Since application of FV method has been discussed in Chapter 3, here I will just discuss about the computation model of the BiAGNR system.

Theoretical investigation of the vibrational properties of BiAGNR has been carried out utilizing the FV method. The width of the monolayer armchair graphene nanoribbons (MAGNR) is specified by the number of dimmer lines throughout the ribbon span [108-112], as depicted in Figurre 4.1. This phenomenon can be expressed by the following equation-

$$S = \frac{1}{2}(N - 1)\alpha \qquad\qquad \text{Equation 4.1}$$

where S is the span of the ribbon, N denotes the number of dimmer lines and α is the lattice constant. Throughout my work, α=0.246 nm is considered by acknowledging the fact that the average bond length of MAGNR coincides with the average bond length of a softened graphene sheet. Taking the free boundary conditions (FBCs) into account, the width of 22-MAGNR is restricted to a finite value of ~2.6 nm while the length is ~30nm encompassing approximately 6900 atoms. Finally, the computational model comprises two such MAGNR to form BiAGNR according to AB stacking configuration. I induce vacancy into the BiAGNR honeycomb lattice system by applying the site percolation scheme. Fig. 4.2 shows the schematic of BiAGNR with random vacancies induced in upper, lower, and both layers. Usually, the percolation theory depicts that if the site percolation network of the graphene-like system has a 70 % bond probability, then the system will encompass 30% vacancy defects. Thus, a high density of vacancy defects up to ~30% is used in this work. Gao et al. showed that graphene produced by the chemical reduction of graphene oxide contains a high vacancy concentration of ~9% [75]. According to Schniepp et al. [76], even higher vacancy concentration of ~30% might exist for graphene oxide. Additionally, molecular dynamics simulation demonstrates that accumulation of irradiation-induced damage in the graphene membrane with vacancies up to ~35% may not show any sign of structural breakdown [77].

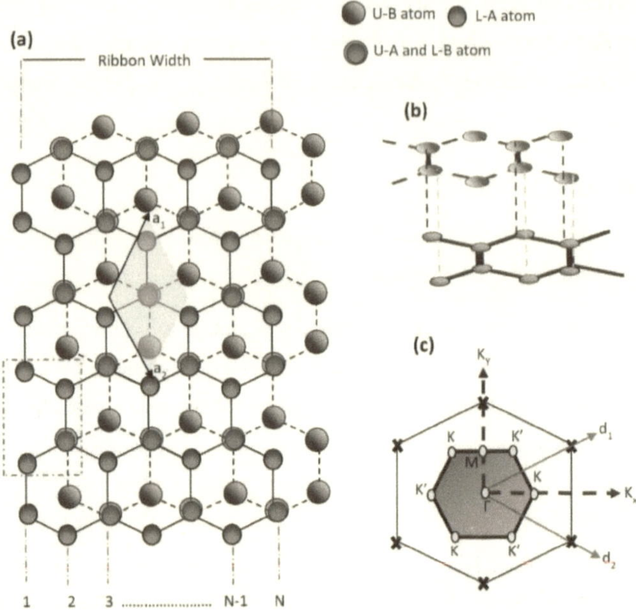

Fig. 4.1. (a) Top view of A-B stacked N-BiAGNR lattice. The width of the ribbon is represented by the dimmer lines N. The shaded parallelogram denotes the unit cell of SLG, where a_1 and a_2 are primitive lattice vectors. U-A and U-B denote upper layer A and B type atoms, respectively. L-A and L-B represent lower layer A and B type atoms, respectively. (b) Side view of A-B stacked bilayer graphene system. (c) A reciprocal lattice of SLG, where crosses represent the lattice points and d_1 and d_2 are reciprocal lattice vectors. The blue-colored hexagon is the first Brillouin zone of graphene with high symmetry points Γ, K, M.

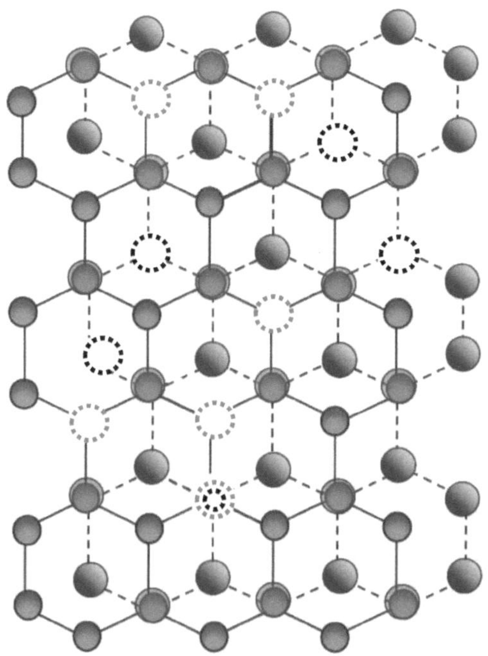

Fig. 4.2- Randomly induced vacancy disordered BiAGNR structure. Red and black dashed circles denote the vacancies in lower and upper layers, respectively. The combined red and black dashed circles represent the vacancies in both layers at the same point.

4.3 Results and Discussion

Fig. 4.3 illustrates the ribbon width dependency on the PDOS of BiAGNR. I have varied the ribbon width over an ample range up to few nanometers and computed the PDOS for individual N-BiAGNR of varying width by FV method. According to Lam et al. [113], N-BiAGNR with a

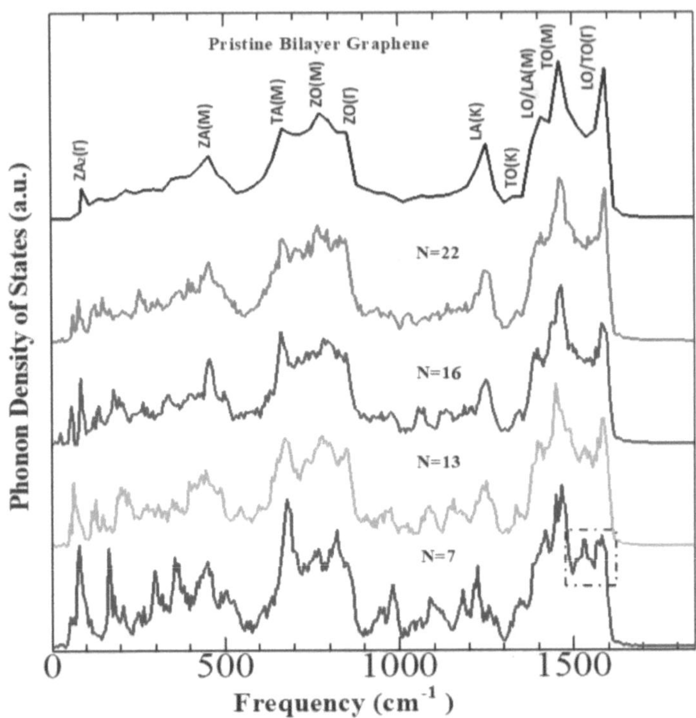

Fig. 4.3. Width dependent PDOS of pristine BiAGNR. Phonon peaks related to different phonon branches at K, M and Γ-points of BZ were defined for pristine bilayer graphene sheet (black solid line).

width less than 2.4 nm possesses a higher energy gap for N=3p+1 than N=3p, where p is an integer; hence I consider the value of N=3p+1 only. It is obvious that six phonon branches are present in any pristine single-layer graphene; 6n branches for n layer graphene. Nika et. al. found a low-frequency peak in the PDOS of bilayer graphene at ~100 cm⁻¹ [114], which is not present in single-layer graphene. That low-frequency regime peak is relevant to Γ-point ZA₂ mode phonons in bilayer graphene. However, for narrower ribbons, peaks related to overtone phonon modes can be found in addition to those fundamental modes of pristine graphene system.

The computed PDOS reveals that Γ-point out of plane acoustic (ZA_2) peak has been observed at ~100 cm^{-1}, which is unique to bilayer graphene. Overtones of fundamental modes are also clearly detected in the PDOS of narrower ribbons. For instance, an extra peak at ~1540 cm^{-1} as indicated by the dashed square in Fig. 4.3 was found for 7-BiAGNR. Moreover, the PDOS become malformed with the reduction of ribbon width. Conversely, when the ribbon width is extended, major peaks associated with fundamental phonon modes are observed in the PDOS and for a sufficiently wide ribbon (22-BiAGNR of width 2.6 nm), the PDOS conforms with the structure of the bilayer graphene sheet.

Among the phonon modes, Γ-point transverse optical (TO) and longitudinal optical (LO) phonons are doubly degenerate, which is also known as E_{2g} mode, and responsible for typical Raman G peak at ~1590 cm^{-1} in graphene [27, 115]. However, the breaking of LO-TO modes degeneracy is exposed for the nanoribbon and LO phonons are found at lower energies than TO phonons [88]. Besides, in armchair-edged graphene nanoribbon (AGNR), only LO mode is Raman active and compared with the E_{2g} mode of pristine bilayer graphene, the Raman spectrum near the edge shows a downshift of this mode. The eigenvector associated with the Γ-point LO (TO) mode phonons are parallel (perpendicular) to the edges of AGNR. Thus, in the Raman process, Γ-point LO phonon can couple to the electron-hole pair created by vertical transition which reasons a downshift in LO mode [116]. Nonetheless, TO mode remains decoupled from electron-hole pair and become insensitive to resonant Raman spectroscopy. Therefore, TO mode in AGNR stay identical to the degenerate Γ-point LO-TO mode frequency (~1590 cm^{-1}) of the graphene sheet. The computed PDOS also reveals a downward shift of the LO mode peak for reduced ribbon width as illustrated in Fig. 4.3.

Furthermore, K-point iTO mode phonons have a greater impact on triggering D mode at ~1350 cm^{-1} in doubly resonance Raman spectrum at the edges of any armchair edge GNRs [33, 115]. In any AGNR system, electronic momenta are perpendicular to the edge of the ribbon. Therefore, being localized around the edges, electrons and holes can recombine radiatively at K-point Dirac cone and the total exchanged momenta satisfies the intervalley scattering between K and neighboring K' point which triggers the D band in doubly resonance Raman spectrum at the edges of AGNR. From Fig. 4.3 it can be pointed out that, peak relevant to K-point iTO phonons arises at

~1350 cm^{-1} for narrower ribbons, though with the increase of ribbon width this peak becomes flattened and conforms with a bilayer graphene sheet.

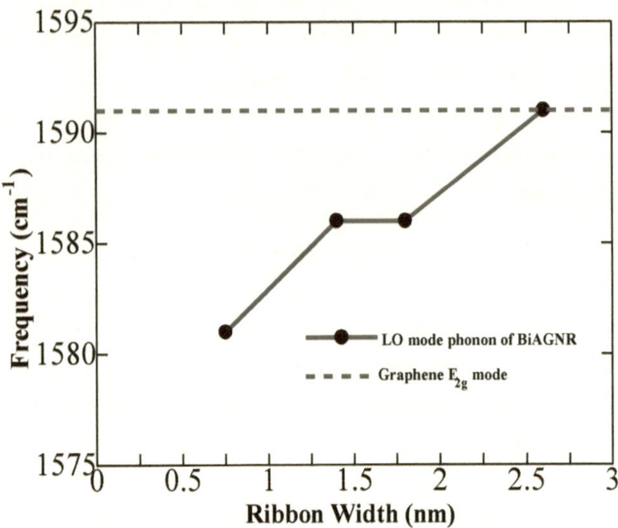

Fig. 4.4. LO mode frequency as a function of ribbon width.

I have then compared the width dependent LO phonon frequencies with the E$_{2g}$ mode frequency of BLG sheet computed by the FV method. As demonstrated in Fig. 4.4, LO mode frequencies have an upward trend with the broadening ribbon width and intersect with the E$_{2g}$ mode frequency (1591 cm^{-1}) while the width of the ribbon extends to 2.6 nm. Under FBCs, the force constants of center region atoms of BiAGNR are well above than edge atoms, which might cause nondegenerate phonons and downshifted the LO phonon frequency of BiAGNR system. It can be noted here that, the downshifting of LO phonon frequency of BiAGNR system is much less than the MAGNR system of the same width as investigated in previous works [93, 117], for which interlayer interactions might be responsible, which employ larger force constants in edge atoms of BiAGNR than MAGNR system.

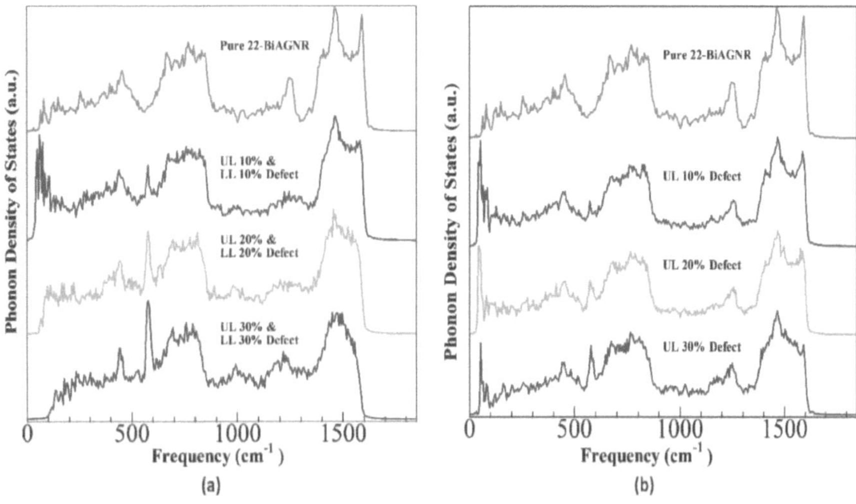

Fig. 4.5. PDOS of 22-BiAGNR with various concentrations of vacancy defects in (a) both layers and (b) upper layer. UL and LL represent the upper and lower layer, respectively.

Apart from the width dependency, the PDOS of BiAGNR also depends on the concentration of vacancy induced in the system. To investigate the vacancy effect on PDOS, I take 22-BiAGNR into account which has a physical width of 2.6 nm. While I consider a pristine 22-BiAGNR system, the PDOS exhibits all typical peaks analogous to any bilayer graphene system [118]. But after inducing vacancies in the system, the LO mode phonon (LO_{mp}) peak of PDOS tends to be demolished as demonstrated in Fig. 4.5; subsequently, the LO_{mp} peak totally disappears when the vacancy concentration is raised to 10% and more. Vacancies are responsible for crumbling the long-range order in the BiAGNR system which acts as a catalyst for the deformation of the LO_{mp} peak. Xie et al. revealed that phonons can be scattered by vacancies in any 2D materials [119]. More phonons are stroked by the vacancies when the vacancy concentration is raised and subsequently the lifetime and mean free path of analogous phonon modes become shortened, thus broadened phonon modes arise. Another interesting phenomenon can be observed from the PDOS of defect induced BiAGNR. A few good-looking peaks arise in the low-frequency regime when vacancies are present, as illustrated in Fig. 4.5(a) and 4.5(b). It can also be observed that, when defects are present in both layers of the 22-BiAGNR lattice, peaks in the low-frequency regime are much higher than the sample in which the same amount of vacancy was induced only in the upper layer. With the increased amount of vacancies, the amount of unsaturated carbon atoms also increases, which is mainly the reason for the lower density in the high-frequency regime. However, a reduction in PDOS at a low-frequency regime can also be observed. This might be caused by low frequency based acoustic flexed phonons. Flexural phonon modes exhibit a quadratic dispersion relation, thus causes an increased amount of low energy phonons in any graphene system [82-83]. Again, vacancy employed lattice specific tension tends to fortify the flexural dispersion, which causes a reduction in PDOS of the vacancy disordered BiAGNR system.

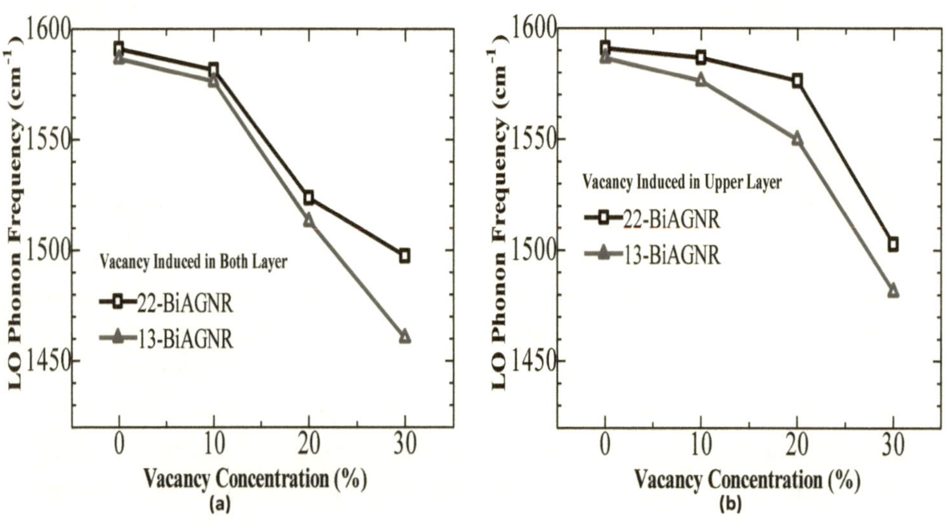

Fig. 4.6 LO mode frequency as a function of vacancy concentration induced in (a) both layers and (b) upper layer.

Vacancy effects on demolishing the LO_{mp} peak can be observed in Fig. 4.6. I have investigated that vacancy has a greater effect on demolishing LO_{mp} peak than the width dependency as illustrated in Fig. 4.4. From Fig. 4.6 it can also be seen that LO_{mp} peak softens more quickly in narrower width ribbon, which suggests that vacancy induced effects are less significant in a wider ribbon. Hereby, electron transport properties are greatly influenced by the vacancy concentration in narrower BiAGNR. Further investigation on Fig. 4.6 also implies that LO_{mp} peak mitigates from the actual position at a higher rate when defects are induced in both layers. Thus, vacancies have a greater effect on the BiAGNR system than the MAGNR.

Phonon modes in BiAGNR lattice can be hindered at the edges and the defect sites due to the symmetry breakdown of the system. Phonons get confined around the edges and vacancies, and localized phonon modes could be observed. Though vacancies and edges both are responsible for localized phonon modes in BiAGNR, vacancies have a greater effect on localization, for which the unsaturated dangling bonds around the vacancy induced sites are responsible. Due to its ruling consequences on the D band properties of Raman spectra of armchair edged and defective BiAGNR, I here anticipate the eigenmodes of K-point in-plane TO (iTO) phonons at 1355 cm^{-1}. D band of Raman spectra is intensely activated in armchair edge BLGNR, for which the intervalley scattering between K and neighboring K′ points is responsible, which occurs due to interchanged momentum from ideal armchair edges. Conversely, in zigzag edge BLGNR interchanged momentum is not able to connect Dirac cones between K and neighboring K′ point, thus cannot satisfy the double resonance condition.

I calculate the mode pattern for 13-BiAGNR with 700 atoms by considering the fact that, modes are intensely localized within 700 atoms based lattice system. A comparison of the localization phenomenon between the edge and the normal phonon modes has also been made. Fig. 4.7 shows the mode patterns of the pristine 13-BiAGNR, 22-BiAGNR, and 24-BiAGNR systems, where all circular shapes are considered as atoms while the colors specify the displacement. I choose a ribbon width of 24-BiAGNR (i.e., ~2.8 nm), which is slightly higher than 2.6 nm (i.e., 22-BiAGNR) to show the differences between the extended and localized states. In narrower nanoribbon, for both upper and lower layers, it can be noticed that all phonon modes are localized encircling the armchair edges (Fig. 4.7(a) and 4.7(b)). This phenomenon conceptually correspondences with an intense D band peak of Raman spectra of armchair GNRs. However, the eigenmodes are scattered within the edges and inside the ribbon with the increase of width, as illustrated in Fig. 4.7(c). This indicates that the phonon eigenmodes are extended throughout the whole system if the width of the ribbon is higher than that of the 22-BiAGNR.

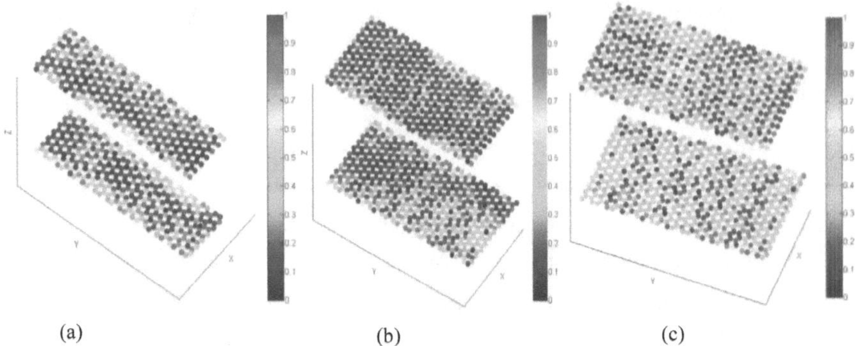

(a)   (b)   (c)

Fig. 4.7- Mode patterns of pristine (a) 13-BiAGNR, (b) 22-BiAGNR, and (c) 24-BiAGNR.

Furthermore, I investigate the effects of vacancies on the localized modes of BiAGNR system. Fig. 4.8 demonstrates the uneven distribution of localized modes encircling the vacancies and the edges of BiAGNR. It can be observed that, with more induced vacancies, phonon modes tend to be localized alongside with vacancies only. At first, I compute phonon modes of the sample where vacancies are present only in the upper layer, as illustrated in Fig. 4.8(a)-(c). Then defects were induced in both layers, as shown in Fig. 4.8(d)-(f). By comparing Fig. 4.8(a) with Fig. 4.8 (b) or 4.8(d) with Fig. 4.8(e), I notice that, when vacancy concentration increases from 10% to 20%, either only in the upper layer or in both layers, more phonon modes are localized around vacancies rather than edges. The strongly localized mode of K point iTO phonons due to the vacancies in BiAGNRs is conceptually well agreed with the large D band peak in the defective GNRs system. It can also be noted that only a few modes are strongly localized which might be caused by the resonance of arbitrarily distributed atoms in the percolation network of the BiAGNR model. However, after a long time, the locations of the central site of localized eigenmodes are shifted from the former locations. In the employed FV method, an outward arbitrary force was applied to compute the mode patterns, which correspondences with the displacements of atoms in every time-varying step, thus it can be assumed that eigenmodes' position is changed with the time. Again, localized eigenmodes encircled the armchair edges in the pristine BiAGNR system, however, their center can be distributed unevenly around the vacancies when few defects are present in the system. If more vacancies are present, the localized modes will encircle the defective sites symmetrically.

46

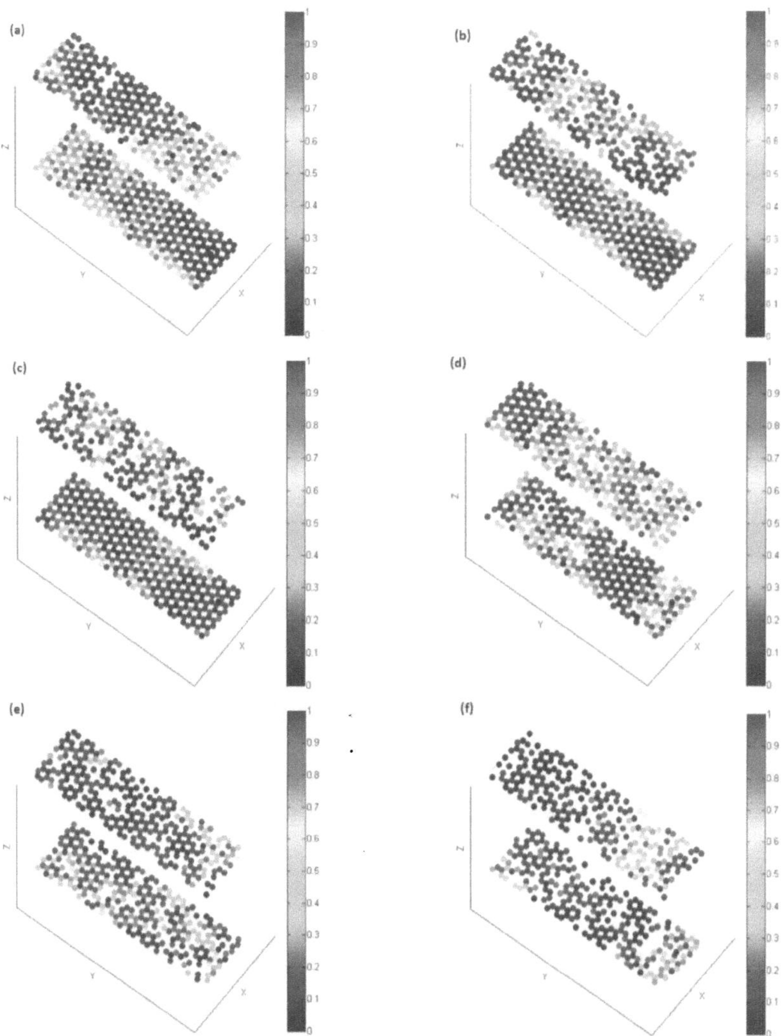

Fig. 4.8. Mode patterns of 13-BiAGNR with different defect density (a)-(c) upper layer 10-30%
and (d)-(f) both layers 10-30%.

It is well known that phonon transport in any system can be stuck by defects. To be more specific, in any one-dimensional system phonons might be localized, while the localization effect increases at an exponential rate in any 2D system. For the 3D system, defects force extended states to alter into localized states. Neighboring atoms of any defective site face intense vibration. This high amplitude vibration deteriorates with the distance from the defective site. This characteristic distance is known as the localization length. For computing localization length, I first calculated the inverse participation ratio (IPR), which can be expressed as

$$IPR_\lambda = \frac{\sum_{l=1}^{N} |u_{l,\lambda}|^4}{(\sum_{l=1}^{N} |u_{l,\lambda}|^2)^2} \qquad\qquad \text{Equation 4.6}$$

where $u_{l,\lambda}$ is the displacement of lth atomic site in λ eigenmode. IPR_λ is a varying quantity that fluctuates gradually between $1/N$ to 1 from extended states to certain localized states. Sheng et al. showed that the localization length of an individual eigenmode λ is inversely proportional to IPR as $\zeta_\lambda \propto IPR_\lambda^{-1/2}$ [120]. The localization length of eigenstate λ can be computed as

$$\frac{\zeta_\lambda}{\zeta_0} = \left(\frac{IPR_0}{IPR_\lambda}\right)^{1/2} \qquad\qquad \text{Equation 4.7}$$

where IPR_0 is the mean value of IPR_λ for the pristine BiAGNR system and ζ_0 is the size of that system.

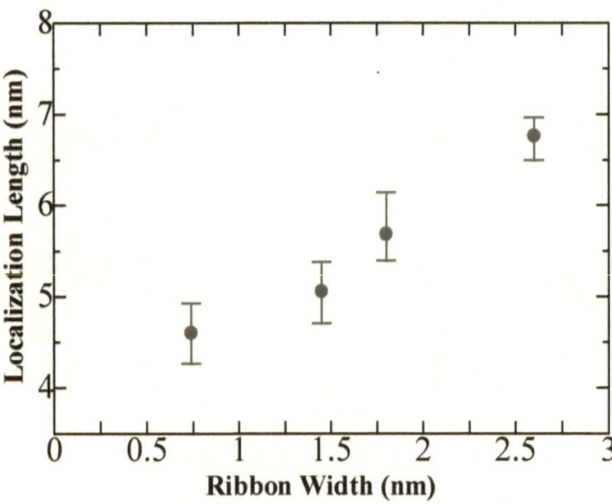

Fig. 4.9. Ribbon width dependent phonon localization length.

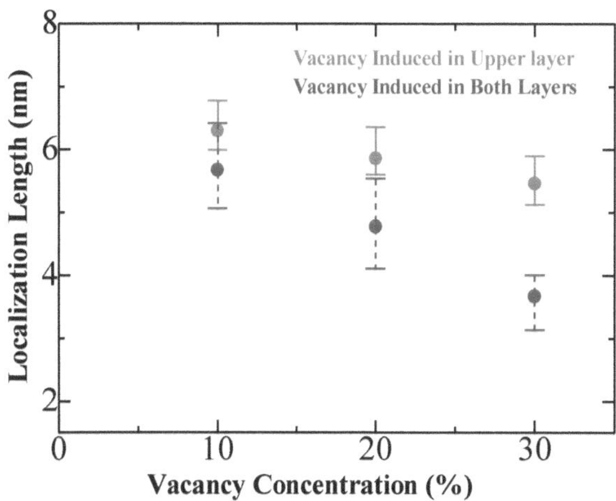

Fig. 4.10. Phonon localization length as a function of vacancy concentration induced in both layers and only in upper layer.

Fig. 4.9 demonstrates the width dependent localization length calculated for the K-point iTO mode phonons at $\omega \approx 1355$ cm^{-1}. The black circle symbolizes the average localization length. Ten eigenstates were taken to compute the average localization length because highly localized sites exhibit larger swings corresponding to the short localization length. As seen in Fig. 4.9 localization length decreases with the narrower ribbon width since phonons around localized sites get the more compact region to propagate in the narrower ribbon. Thus, highly localized states are observed in the narrower ribbon. In Fig. 4.10, I express the localization length by means of vacancy concentration. I first induce vacancies in the upper layer and then in both layers of the 22-BiAGNR. I have varied the vacancy concentration from 10% to 30% in both cases. With the increased amount of vacancies in the upper layer, the localization length tends to decrease, which signifies that higher vacancy concentration develops higher localized states. Moreover, I can also observe from Fig. 4.10 that, when I induce vacancies in both layers, for the same amount of vacancies the localization length decreases much as compared to the case where only the upper layer was considered.

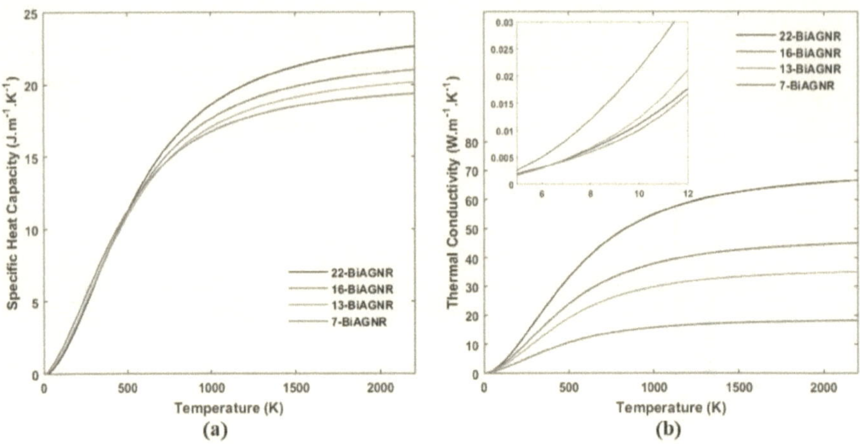

Fig. 4.11. Width dependent (a) specific heat capacity and (b) thermal conductivity of BiAGNR system.

Finally, I have employed quantum statistical theory [121], to compute the specific heat capacity of phonons (C_p) in BiAGNRs. Unlike the continuum-medium model, my computation is not only limited to the low-temperature regime [122]. From the computed values of the specific heat capacity, I have also quantified the thermal conductivity due to phonons of BiAGNR system. I can compute thermal conductivity of phonons as- $\sigma_p = \sum C_p . v . \lambda$ [120], where v is the group velocity of phonons [123-124], and λ is the mean free path of phonons in BiAGNR system which might be assumed to be $\geq$W for nanoribbon while W is the width of the ribbon. In Fig. 4.11, specific heat capacity and thermal conductivity of BiAGNR have been shown as a function of ribbon width. Fig. 4.11(a) depicts that at the low-temperature region, specific heat capacity is greater in the narrower ribbon width, which proves the highly localized states along the edges in the narrower ribbon. With the increase of temperature, the specific heat capacity of the narrower ribbon tends to decrease since at higher temperatures, the density of excited phonon modes in the wider ribbon becomes much higher than the narrower ribbon. Near the edge of the BiAGNR, lattice vibration is far more different than the central carbon atoms due to the effect of symmetry lowering. From my previous discussion of Fig. 4.3, abrupt changes in PDOS can also be observed with narrowing the ribbon width which in turn depicts the width dependent specific heat capacity of BiAGNRs. Moreover, it can be noticed that, for 22-BiAGNR system, when T>2000 K, the specific heat capacity tends to become identical with increasing temperature since for any graphene system, the in-plane Debye temperature is ~2200K and specific heat capacity of BiAGNR reaches a value of ~22 $Jm^{-1}K^{-1}$ which is less than the classical Dulong-Petit limit of graphene system (25 $Jm^{-1}K^{-1}$) [125-126], as it should be for graphene nanoribbon system.

Fig. 4.11(b) illustrates the temperature-dependent thermal conductivity of BiAGNRs for different ribbon width. We can observe that the thermal conductivity is higher for the wider ribbon at the high-temperature regime. It is well known that the thermal conductivity of nanoribbons due to phonons depends on W^{β}, where W is the width of the ribbon and β is a factor which changes in each case while phonon transport is localized, diffusive or ballistic. Generally, $\beta \gg 1$ for localized phonons, $\beta=1$ for diffusive phonons and $\beta<1$ for semi-ballistic phonons [127]. Due to highly localized states at very low temperatures, the thermal conductivity of the narrower ribbon tends to rise higher since β is higher for the narrower ribbon. With the increase of temperature, phonons get excited and localization becomes minimized even in narrower ribbon, which turns $\beta \approx 1$. Hence, thermal conductivity becomes identical at a higher temperature (T>2000K) for a certain width. The thermal conductivity of 22-BiAGNR (width~ 2.6nm and length ~30nm,) reaches a value of ~68 $Wm^{-1}K^{-1}$, which is fairly analogous to previous studies [128-133]. Though a wider ribbon shows higher thermal conductivity at the high-temperature region if we zoom in Fig. 4.11(b), we can see that at a very low temperature the narrower ribbon shows higher thermal conductivity which is illustrated in the inset of Fig. 4.11(b) This suggests that β becomes negative at a very low temperature which is caused by low-frequency regime acoustic phonon modes, which has also been studied in previous literature [134]. The low-frequency regime acoustic phonons do not depend on the width of ribbon-like the higher frequency regime phonon modes do and thus they are hardly affected by the edge roughness of BiAGNR system [127].

I have also computed the specific heat capacity and thermal conductivity as a function of different vacancy concentration as illustrated in Fig. 4.12. From Fig. 4.12(a), it can be observed that with increasing vacancy concentration the specific heat capacity decreases due to the combined effect of edge and vacancies in BiAGNR. I also investigate that when defects are induced in both layers of the 22-BiAGNR system, it shows lower C_p values than the one when defects are induced only in the upper layer.

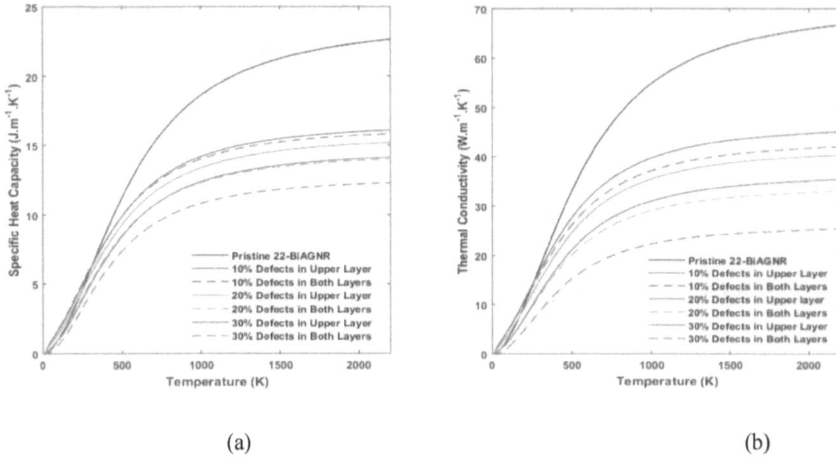

(a)   (b)

4.12. (a) Specific heat capacity and (b) thermal conductivity of 22-BiAGNR with different vacancy concentration.

Furthermore, it can be observed that, with high vacancy concentration, specific heat capacity doesn't change hastily. In spite of phonon being localized around the vacant sites, the center of localized modes advances to vacant sites with time development, which in turn depicts that heat conduction does not depend on the localized phonon modes near vacancies, but due to the non-defective part of the BiAGNR system. Fig. 4.12(b) represents the thermal conductivity with respect to different vacancy concentration. Since at higher temperature, the phonons at the non-defective region get much excited, the thermal conductivity increases with temperature. However, when vacancies are induced, the thermal conductivity decreases since phonons get localized around the vacant sites which sequentially points that the mean free path of phonons, λ also tends to be decreased at the presence of vacancies.

4.4 Conclusion

In summary, I have studied the vacancy effects on the vibrational properties of BiAGNR using the FV method. Compared to MAGNR, my results show that the width of BiAGNR has less effect on the phonon modes. Apart from the width dependency, the PDOS of BiAGNR strongly depends on the concentration of vacancy induced in the system. I observe that LO mode phonons are shifted downward with the increase of vacancy concentrations, however, the downshifting of the LO frequency in the BiAGNR system is much less than the MAGNR system of the same width. Vacancies persuade some new peaks in the low-frequency region of PDOS. It is found that, when defects are present in both layers of BiAGNR lattice, peak intensities in the low-frequency regime are higher than that of the sample with the same amount of vacancy induced only in the single layer. I also quantify the phonon localization phenomena of disordered BiAGNR. A strongly localized eigenmode is perceived near the edge and vacancy sites in BiAGNR systems. My calculated typical mode patterns elucidate that the K-point iTO phonons are shifted towards the defect sites from the edges with the increase of vacancy concentrations. I further showed that the phonon specific heat capacity and the thermal conductivity significantly decrease with the increase of defect density or the reduction of the width of the ribbon. These findings provide a deep insight into the heat dissipation phenomena of graphene-based high-performance nanodevices and elucidate the Raman as well as experiments related to the phonon properties of BGNRs.

Chapter 5. Conclusion and Future works

5.1 Conclusion of the book

The findings of this book can be summarized as follows-

1. Combined ^{13}C isotope and vacancy effect on the phonon properties of bilayer graphene have been studied. Phonon density of states has been computed by varying ^{13}C isotopes over a wide range possible and also by varying vacancies as rational as possible to a practical scenario. From my computed results, it can be observed that PDOS of bilayer graphene is greatly impacted by the defect concentration and with the increase of defect concentration more broadening and softening of PDOS peaks have been observed. Moreover, lattice vibrations are found to be responsible for the high peaks in the low-frequency regime which is caused by the defect sites in bilayer graphene. Furthermore, a split of Raman active G peak has also been noticed for samples where ^{13}C isotopes have been induced in one layer of the bilayer sample. In addition, the impact of combined isotopes and vacancies on the localization of optical phonons have been analyzed. Mode pattern and localization length have been quantified for K-point in-plane transverse optical mode phonons. Localized modes have been noticed around the defect sites and localization length has been found to its lowest for combined isotope and vacancy disordered samples. After that, localization length has been computed for E_{2g} mode frequency at ~1590 cm-1 for by varying the isotope concentration over an ample range. With the increase of isotope concentration, average localization lengths have found to be decreased at first and then increased.

2. I have studied the vacancy effects on the vibrational properties of armchair edge bilayer graphene nanoribbon by using the FV method. From the comparison with MAGNR, my results show that the width of BiAGNR has less effect on the phonon modes. Moreover, the PDOS of BiAGNR also strongly depends on the concentration of vacancy induced in the system. LO mode phonons have been found to be shifted downward with the increase of vacancy concentrations. Nonetheless, the downshifting of the LO frequency in the BiAGNR system is much less than the MAGNR system of the same width. Some new peaks in the low-frequency region of PDOS have been found which suggest the effect of unsaturated dangling bonds on the

phonon modes at the presence of vacancies. It is found that, when defects are present in both layers of BiAGNR lattice, peak intensities in the low-frequency regime are higher than that of the sample with the same amount of vacancy induced only in the single layer. The phonon localization phenomena of disordered BiAGNR have also been studied. Strongly localized eigenmodes have been found near the edge and vacancy sites in BiAGNR systems. The calculated typical mode patterns elucidate that the K-point iTO phonons are shifted towards the defect sites from the edges with the increase of vacancy concentrations. After that, the phonon specific heat capacity and the thermal conductivity have been computed. A significant decrease in phonon conduction has been observed with the increase of defect density or the reduction of the width of the ribbon.

These findings provide an insight into the heat dissipation phenomena of bilayer graphene-based high-performance nanodevices and also the Raman spectra as well as experiments related to the phonon properties of different bilayer graphene systems. Thus, the FV method can act as a versatile technique to compute the Raman spectrum of any disordered sample.

5.2 Future works

The model employed in the calculation is justified to compute the vibrational properties of 2D bilayer graphene. Though, there are few things that can be done in future; for instance, a study on phonon properties of a disordered layered model of two different materials like graphene and h-BN, new complex 2D materials like MoS_2, WSe_2, silicene, SiC and so on.

This work was done by considering bilayer graphene which will help to model the layered system of graphene and SiC or h-BN. Furthermore, currently, I am working on modeling the disordered silicene system which is also proven to be a potential material for nanoelectronic applications. Moreover, new promising 2D materials like SiC and MoS_2 can also be modeled by this method.

Raman spectroscopy has been proved to be an extraordinary tool to analyze the physical properties of any 2D materials. Most of the simulation work on Raman spectroscopy was done by employing a direct diagonalization technique or moment method; both of the techniques consume a lot of computer memory and need a huge time to compute. However, employing the FV method, Raman intensity can be computed within a reasonable amount of time since it takes lower computation memory. In addition, no computation of Raman intensities has been taken place for any disordered samples. Therefore, future work should be emphasized on employing the FV method to compute the Raman intensities.

www.ingramcontent.com/pod-product-compliance
Lightning Source LLC
LaVergne TN
LVHW041754190726
843493LV00008B/2627